Humans in the Making

Social Interdisciplinarity Set
coordinated by
Georges Guille-Escuret

Volume 4

Humans in the Making

In the Beginning was Technique

Michel J.F. Dubois

iSTE

WILEY

First published 2020 in Great Britain and the United States by ISTE Ltd and John Wiley & Sons, Inc.

ISTE Ltd
27-37 St George's Road
London SW19 4EU
UK

www.iste.co.uk

John Wiley & Sons, Inc.
111 River Street
Hoboken, NJ 07030
USA

www.wiley.com

Library of Congress Control Number: 2020941846

British Library Cataloguing-in-Publication Data
A CIP record for this book is available from the British Library
ISBN 978-1-78630-584-8

MIX
Paper from responsible sources
FSC www.fsc.org FSC® C013604

Contents

Introduction

Human Ontophylogenesis

"The highest philosophical initiation is therefore not only a knowledge of models (Ideas), but a mode of being that makes the philosopher coincide with the absolute source of forms and existences; it is indeed the intuition of anticipation in its purest state [...]. It's not the movement, but the intuition of any projection into existence and multiplicity."

Gilbert Simondon, *Imagination et invention*, Puf, Paris, p. 59.

"Then thus: 'In the Beginning was the *Thought.*' This first line let me weigh completely, let my impatient pen proceed too fleetly. Is it the *Thought* which works, created, indeed? 'In the Beginning was the *Power,*' I read. Yet, as I write, a warning is suggested, that I the sense may not have fairly tested. The Spirit aids me: now I see the light!
'In the Beginning was the *Act,*' I write."

Goethe, *Faust*, I, 3.

"In order to understand what happened in the past, we must interpret the empirical facts without our socio-political ideas interfering. [...] Any paradigm can be misused for ideological purposes, because models, whatever they may be – cognitive, hermeneutic or evolutionary – lead too easily to to positions that have little to do with the reality of the facts and sometimes even deny the existence of some."

Marylène Patou-Mathis, *Préhistoire de la violence et de la guerre*, Odile Jacob, Paris, p. 163.

We shall set out a narrative of the emergence of the human, or humans, for the widest possible audience, maintaining the scientific and philosophical perspective necessary to prevent this narrative from becoming fiction. We have enough scientific and technical knowledge, to date, to consider reconstructing, step by step, the path leading from a distant population of higher primates, about eight million years ago, to humans. To do so, we must make use of the last two centuries of research in the biological and geological sciences, anthropology, paleontology, ethnology, neurology, cognitive sciences and philosophy[1].

This is therefore a heuristic narrative (useful for research) that sets out a number of hypotheses. The aim is to advance understanding rather than to oppose the working hypotheses or conclusions of scientists. These hypotheses are consistent both with the facts gathered and with an evolutionary approach, which itself is capable of integrating them. Lamarckian-type adaptationist approaches will be carefully corrected. Because often, talking about adaptation and not selection is reasoning according to the Lamarckian approach. The fundamental hypothesis defended here remains Darwinian, because it is not a question of acquired traits being hereditary but of learning abilities being selected. We could assess our position as more Wallacean (after Wallace, co-inventor of selection theory) than Darwinian because of the integration of multilevel selections, but these are subtleties of specialists (Gayon and Petit 2019, p. 127–143).

There is no reason to present the emergence of the human as a mystery within the world of living things. There are many surprising living beings and the enigma of the source of life is far from being solved. It is a difficult problem, but its solution requires multidisciplinary research. The origin of humans can be seen as a given; they are part of primates, among which are the great apes. Their lineage separated from that of the genus *Pan* (chimpanzees and bonobos) about seven million years ago, according to both paleontology and molecular biology. From the moment that the genealogy that leads to humans is thus recognized, whose order of changes can be

1 The reader may refer to the chronologies and species of hominids available on the Internet: https://annex.exploratorium.edu/evidence/lowbandwidth/INT_hominid_timeline.html.

identified, we can make well-founded, useful and even suggestive hypotheses to help us know ourselves[2].

An estimation of what is innate to humans cannot be a comparison to other animals without distinction nor evolutionary analysis. For example, there is no reason to compare man to the penguin, the meerkat or to some ancient reptiles that would have been vertical, given that verticality does not come from the same process, or to dolphins with very large brains, since they do not use their intelligence in the same way and do not live in the same environment. The evolution of their living conditions over the last seven million years has been different. The architecture of the brain and its connection to the body is as important as the size. Furthermore, in the analysis of human populations, there is great variability in the size of the brain, without it being possible to draw any conclusions about intelligence. Focusing on the brain is no more relevant than focusing on verticality or socialization. What is most relevant is to have an evolutionary approach to understanding the process of separation from other great apes and to make relevant comparisons where useful and possible.

Speaking of emotions, Darwin was a pioneer in showing that the expression of emotions is visible in most mammals. Many other characteristics, generously attributed to humans alone, are present in primates and often in other mammalian lines: for example, attention span, memory of events, anticipation, laughter, and so on. Infanticide, lack of maternal instinct, fighting to the death/murder, or even "wars" between groups within the species, homosexuality, masturbation, etc. have been perceived as human perversions. They are found to exist in primates and even in other mammals. It is out of ignorance that we have judged our "exception", whether in terms of values or cognitive abilities, when the most relevant differences may lie elsewhere. We are living beings whose general modes of functioning we share, and the closer species are to each other, that is separated from us for a "shorter" time, the greater and more numerous the similarities.

---

2 For clarity of presentation, I try to follow the classification that seems to be the most common: the species that are on the lineages that led to chimpanzees and bonobos are from the genus *Pan,* and can be called panins. The separation to form the various *Australopithecus* and other upright apes are hominins, leading to the genus *Homo.* The two groups (hominins and panins) form the hominids. This separates the hominids (globally most of great apes, including *Homo sapiens*) in two parts according to one criterion: stable bipedalism. Generally speaking, we speak of *homininae* when talking about "species" named *Homo*; the first one being *Homo habilis.*

Descriptions of what makes or would make up the originality of the human being as it appears today, in relation to living things, are not our immediate purpose. The main thesis defended here is the result of the search for what has allowed this originality to exist and to show itself. *It describes the conceivable process of its genesis over a very long period of time, in millions of years, that is to say its phylogenesis, and not how an individual is constructed from conception to maturity, or its ontogenesis.* In doing so, human originality reappears, but in a different way. What is human will not appear as the *unum necessarium* sought after by philosophers and metaphysicians; it may appear to be a combination of characteristics that have co-evolved, sometimes in convergence, sometimes in juxtaposition, under a selection pressure that is quite stable enough that it becomes possible to conceive. The problem becomes one of researching and defining the nature of this selection pressure.

The human being, considered in its many facets, can be described as the greatest predator that ever existed, the first living species seeking to organize the living world rather than to fit into it and the first also to undertake a permanent evolutionary process of transforming its own way of life, and this in spatial and temporal heterogeneity, even if it means disrupting all ecosystems and living beings. Many of its members believe that they are descendants of an original created couple and are therefore radically different from other living beings, even the closest ones, and sometimes even from other human groups. For the human being also seems to be the only living being that invents worlds in order to fit into them and bring other living beings into them.

It is the only large species that are both unsuitable and present almost everywhere, thanks to technique. Creating, inventing, innovating: it all started a few million years ago, alternating apparent "stasis" with periods of more rapid transformation modifying pre-human and then human conditions of existence. Change accelerated during this process. We see the planetary consequences of this incessant expansion. An invasive species, if ever there was one; extra-terrestrials landing on earth would identify it immediately as it occupies space in many ways, different in each place. This inventiveness is constantly increasing; now it is engaged in understanding what has allowed it to appear and in modifying the rules of the game by which it entered the world of the living.

Our analysis attempts to unravel the enigma of this specific process concerning groups of great apes that have gradually, over hundreds of thousands of generations, departed from their initial conditions of existence: a more or less peaceful life in the African rainforest or tropical rainforest. Step by step, they left this environment or biotope until they reached the scale of a global phenomenon. Jared Diamond (2001) considers humans to be "the third chimpanzee" because we are so close to chimpanzees and bonobos. Based on the qualities of the "other two chimpanzees", can we understand how this divergence which would lead to humans occurred, which has become so spectacular on a global scale?

We will use the term *human* in different senses. There's the name given to the species. Here, we speak of humans and not of man, because in French (but also in English) "*homme*" is gendered, and it now seems preferable to speak of a human (here neutral) who can be a man or a woman. In French, "*humain*" is also an adjective that has a double meaning, because it is opposed to "inhuman", although the latter is human. It is therefore perceived, at least in French, that it is possible for a human to be more or less human. With reference to Nietzsche's book (translated into English: *Human, All Too Human*), there can be a scale of the human being, which phylogenesis – that is to say, genesis in the duration and long succession of generations – could account for.

It is not possible to think of humans as a whole without thinking at the same time of the individual human, the human in groups, the human in terms of population and finally as a general, virtual entity that crosses populations. This comes from the fact – and this narrative will attempt to demonstrate this – that the human in its totality, the "human phenomenon" as Teilhard de Chardin put it, goes beyond the individual human. Each individual is human in the sense that he or she refers, consciously or not, to this global human. But no human individual can claim to embody the totality of what the human represents, expresses and manifests. Each of us is conscious of belonging to the human community, but none of us can claim to embody the totality of what it means to be human. Just as there is no human without technique, there is no human without human community. Even a human community, however large, cannot claim to represent the human in the totality of his or her potential. This emergence therefore describes a human being who is always in the making.

PART 1

Phylogenetics of the Emergence of Humans

1

The Long and Slow Emergence of Humans

1.1. The difficulty of thinking about the beginning of the human being

Before examining how this almost magical animal – other people, like J.F. Dortier (2004), would say "strange" – that is, the human being, I think it is important to intuitively grasp the crux of the problem when we try to define it. Who are we? What are we? Where are we from? It seems that as soon as human groups have been able to transcribe, or retain in memory, their conception of the world, they present an origin of the world and of human beings (Godelier 2010). Science is, by its existence and its ceaseless progress, the tangible proof of the possibility of conceiving the world, and of conceiving it in such a way that it is possible to act in it, without seeking to be in line with common sense. Nevertheless, there is – in fact, but also in principle – a major problem when it comes to conceiving the relationship we have with what we conceive. This problem becomes greater when it comes to thinking about our own genealogy.

The objective reality of the world is a set of data that is accessible to us, despite all the mental constructions invented to make us believe the opposite, but it requires a real intellectual effort which is never completed. Setting aside current philosophical discussions that require a "technicality" that one must have learned to understand its substance, we can grasp what objectivity is by drawing on some examples. How is it possible to send men to the moon, or a robot to Mars, if the tools to do so are not derived from concepts and objective methods? How is it possible to make films, and watch them, if what we feel is not accessing reality? I remember ancient discussions with African elders who questioned Western knowledge, blaming it for being what we would translate today as "ethnocentric". Yet they concluded: "There is, despite everything, something true in your knowledge, for an airplane, for example, which can carry hundreds of people in the air, over thousands of kilometres, is proof beyond doubt. But is that so important?" Their wish was that their people could acquire the "objective" knowledge they did not have, while retaining the knowledge they had and that we no longer possess. This is another story that we will not go into here.

What is this "magic" that is unique to humans? It's the fact that only the human knows how to make planes and rockets, and send animals, plants and humans into space. In fact, it all started a long time ago with the production of stone tools. An effort to abstract and question our "natural" thinking about things is necessary, along with a minimum of uncontrollable intuition to try to access this slow and long process that eventually led to the emergence of humans.

Since the emergence that makes us human and being human, we are – and I say "we" thinking of the collective of all humans, since the human emerged, between 300,000 years ago and 50,000 years ago, to which I am aware of belonging – beings conscious of the world: that is, conscious both of the exteriority of the world in which we are and to which we belong, and conscious of our presence in this world, as part of this world; we are a question for ourselves. We are a mobile consciousness, albeit unstable, capable of reflecting on ourselves, capable of projecting ourselves on the world and on others, but also capable of being moved by the encounter with the other who shows us that he too is conscious of himself, projective consciousness, mobile consciousness, conscious of himself and of the other. We are. We are there. But what are we? And how did we appear? As Heidegger (1958) would say, what is this being whose being is such that the question of being has a meaning for themself? Or again, like Sartre (1943), what is this being for which the question of being is in its being involved with another being other than himself? Everyone perceives themselves being; perceives themselves being there; and yet elsewhere, from this elsewhere that allows him to say himself; to say he is there, in the being. Each indvidual knows himself to be a whole, but a whole that knows that he is not the whole at all, because aside from his whole, there is not nothing; he lives it unceasingly.

I have used the "we" here to indicate that when thinking of individual consciousness, the use of "I", separate from the original relationship that allowed it to appear, takes away its being. The relationship, as Gilbert Simondon (2020, pp. 24–27) showed, has a "rank of being". More precisely, as Paul-Antoine Miquel's (2015) analysis demonstrated, it is first in relation to the being (i.e. nature) of the being (i.e. individual) that emerges in relation. The "I" does not appear from nothingness; it is an "awareness", that is, consciousness is what takes and is taken at the time; self-awareness is and is not the same "thing" as the conscious self whose totality the self-awareness believes it takes. "I am conscious of myself" is a displacement of the self in relation to the self which shows, in fact, that when we speak of consciousness, we find ourselves in a world where the principle of the excluded third party no longer works. There is no A or non-A ("conscious of" or "not conscious of"); there can also be A and non-A; and even neither A nor non-A; because the whole of A and non-A is not everything... Within this framework, any logical analysis of the discourse will show a contradiction. This contradiction, which is real, makes the being of the conscious "I" an ambiguous being, at least for the classical logician. Insofar as consciousness is not understood as a process, nor is it possible to reduce it by pure logic, logical analysis cannot grasp it.

The "fact of consciousness" is a logical paradox; we must concede that everything that concerns it is marked by the logic of the "third party included". There is no way out of it, except to look for the reference of what is considered external to consciousness within itself, which is contradictory. An objectification approach refers the object under study in terms of what it is related to, outside the conscience; but this external reference can be internalized within the consciousness, while having been produced outside it. The filmmaker, who captures a scene, can be filmed, while the second filmmaker filming the scene is the one that says the meaning of the film, but whose value is doubled, repositioned, by the one who shows its construction. We could imagine a third filmmaker who would himself be in the first scene and would show in places that it is indeed a scene already observed by the other two, from outside the scene.

If the plane or spacecraft is the indicator of the objective efficiency provided by the scientific and technical approach, we could say that cinema is the indicator of subjective efficiency. As we can see in this last example, tools are needed to create this separation between the observing subject and the observed object, whether this tool is simply a mediator or a third-party mediator that produces a new median space between the object, the first tool, and the user of the latter. The use of tools makes us aware that consciousness is a paradox; it is both a tool and a user of tools, because in essence it is a "being of relation". What is significant is that the more recent the tools are, the more they speak to us about what makes us human. We will come back to this aspect after our long journey in the past.

This paradox of being conscious was experienced by humans at that distant time when they were fully developed. They have encountered a world of consciousness that cannot be understood only by a simple logic excluding the third party: the consciousness of the world, of other conscious people and of oneself, the one that allows access to knowledge and belief, that makes one wise or "crazy", or both at the same time. The first human groups that reached this consciousness of self and others, in a dangerous natural world where the refusal of the excluded third party leads to certain death (is there or is there not a lion; is there or is there not danger; the third party must be excluded!) must have lived it as a terrifying, magical and marvellous experience. It seems that the oldest religions can be called animist (Godelier 2010). Indeed, appearing in the natural world in which human objects remained a minority, this relational consciousness, if it could be said to be between humans, should lead to the belief that every living being, and therefore every being identified with a living being, should also have a similar consciousness. We have learned so much to separate thought from the body that it does not occur to us that we feel more than we think. In ancient times, everyone knew this intuitively, even if they had difficulty saying it. It is probable that in the first moments of self awareness, humans realized that they were not acting in consciousness, but in participation, in perception. Self-awareness came later. This is still the case, but who is aware of it today where self-affirmation is sanctified?

What seems more and more likely is that this "awareness" took place in stages, over a longer period than we think. Could it be the development of technical and then scientific thought and of "established" religions that put it aside? We are at a time of scientific and technical progress (see the above example of cinema, to which the techniques of the virtual are added) such that this problem is resurfacing and can no longer be shrugged off.

1.2. The current challenge of human construction

Without being under any illusion about our ability to explain, in its strongest sense, how this emergence was possible, is there a way to understand it and to track its appearance in the scientific descriptions of the developments that took place over the course of about seven million years? From the ancient great ape, which is our shared ancestor with the chimpanzee and the bonobo, how have the populations that gradually allowed this new form of life, the one we call human, to evolve? Is this new form strictly speaking unheard of because it is based on the possibility of a kind of endless transformation?

Over the past 50,000 years, populations from this "human form" have spread across the globe, multiplying in numbers unimaginable for an animal species of this weight. Today, the human species weighs 350 million tons directly, more than double if we add domesticated animals. We have to multiply that by more than six if we count domesticated plants. Although this represents little more than 0.2% of the total mass of living beings, including microorganisms, which take up half of it in terms of flow, humanity takes up more than a quarter of the primary biological production of the continents, which comes from the photosynthesis activity of plants (Krausmann *et al.* 2013). The inanimate material aspect, that is the minerals and fossilised materials of all kinds that humans exploit, leads to a tenfold increase in the masses involved. Which species takes up so much space?

About 50,000 to 60,000 years after the emergence of "modern" humans, an extremely short time on the Earth's timescale, or even of life on Earth, the human population has taken a unique place in the history of living things. On the basis of a historical analysis, we can claim that the "anthropocene event", this novelty that shakes us, comes first and foremost from the West, and especially from the English-speaking world, or even from capitalism, and that it is very recent (Bonneuil and Fressoz 2017). This is where we confuse how it came about and the overall long-term consequence. Certainly, this "humanity" is not homogeneous. Certainly, some people "weigh" per individual ten times more than others; the fact remains that the trend is a global drive. The West no longer accounts for the majority of the global drain, and its share is declining every year. According to this historical logic, we could say that the beginning of human groups is in Africa, a contingent, specific and local event; it is therefore Africans, from the very beginning, who are "guilty" of the current deregulation, and even just a small population somewhere in Africa... But today, humanity is on every continent, and the pressure of its burden, albeit very unequal, is visible on every continent and in the sea. So, it is not the way a process, an activity, a value is set up that defines the importance of the event. Or again, the reasons for the appearance of a phenomenon do not in any way prejudge its importance in the long term. This is in accordance with the concept of exaptation, which we will discuss later, and which Nietzsche, the inventor of genealogical analysis, had already stressed many times.

Today, understanding the human being is becoming a crucial issue, both intellectually and practically. We cannot be content with saying that human cognition is unequalled (it still has some very archaic aspects) nor with thinking about the human being in this space recently named Gaia – which both understands the human being, interacts with him or her and becomes a condition for the continuation of this experience – as unique since the advent of the first living being. What is the relationship between humans and their environment?

2

Technique and Becoming Human

2.1. A general definition of technique

Technique can reveal the human to himself. Nowadays, the tendency is to qualify any technique that has been developed on the basis of scientific knowledge as technology. This is not the place to discuss the difference between technique and technology, nor to show that the study of technique is also, and was historically, called technology. However, the use of technique predates the birth of science, and even goes beyond the history of the biologically modern human species in its broadest sense. It is found sporadically in some animals, in mammals (including primates) and in birds; it is observed in some arthropods, such as ants, bees or termites; and Darwin also showed something similar in earthworms. But a transformation appears in the homininae lineage, that is to say that the use of many techniques becomes permanent, changing and necessary for survival.

Postmodern techniques which appeared at the end of the 20th Century following the two industrial revolutions are much more complex and their integration into the human world cannot be compared to the use of "prototechnics" by animals, however advanced they may be. The capacity of modern technical systems to intervene in humans themselves is emerging, to the extent that it may become important to re-examine, at a new cost, the question of the human being in terms of their relationship with technique by returning to the first interactions, if it is possible to question them. Before even tackling today's techniques, can we better understand our world by understanding how the technique has become entangled in human life?

There are many definitions of technique. Etymologically, it refers to material production or manufacture. According to Plato, the *techné* (τέχνη, know-how) is in contrast to the *épistémè* (ἐπιστήμη, knowledge); indeed, it is not the application of theoretical knowledge, but practical knowledge that has a link with pedagogy and the arts. Among the ancient Greeks, knowledge did not have the position of domination that it has today over technique, and technique was not thought of as an application of knowledge. By caricaturing a little, the history of philosophy is confused with the desire to impose knowledge to the detriment of technique and consequently to abandon reflection on technique. The revival of thinking about technique began towards the end of the 18th Century, but without much success. Although Benjamin Franklin said that man was a *tool-making animal*, it was only from Bergson, his notion of *Homo faber* and the demonstration of intelligence in the ability to make tools and modify one's environment, that a slow intellectual restructuring of technique was set in motion, in parallel with an unprecedented technical expansion.

The simplest and most obvious definition of that technique is a coordinated set of means and methods to achieve an end. However, many techniques can be one element in more complex heterogeneous sets, to the extent that the end of each technique is no longer intrinsic to it and may depend on the context. Moreover, it is often possible to take several technical paths to the same end. It can be seen that purpose is rarely the driving force behind a technical retouch or the emergence of a technique. More often than not, it is in the aftermath that the purpose shows itself. Likewise, inventions seemed uninteresting when they first appeared, only to undergo important developments a few decades later; for example, the modern laser or the personal computer.

A second approach is to define it as a set of methods that change the conditions of human existence. In fact, the invention and adoption of a new technique can, to a greater or lesser extent, change the conditions of existence of a human group: adoption of a more efficient method of hunting or fishing, the invention of agriculture, metal extraction and metallurgy, the invention of machines, automobiles, smartphones, etc. This revision of the conditions of existence can lead to a reconfiguration of the purposes to the point of transforming the human mode of existence. For there to be technique, it is considered that "technical objects" are needed, classified in three groups: tools, instruments or utensils.

In a broader sense, technique can do without tools. For example, the high jump technique was transformed by Richard Fosbury in 1968, at the Mexico Olympic Games, without external technical input or tools. There are many different techniques of high jumping that depend on the surrounding socio-technical conditions and can be entirely bodily. Likewise, an acrobat, a cliff climber or a dancer can use tools; in some cases, they use almost none and the difference lies in the technical excellence of the body itself. Marcel Mauss (1936) is the originator of the concept of "body techniques". His project was to articulate the mechanical and physiological with the psychological and sociological. We consider here that phylogenetic analysis is a prerequisite to understanding how this articulation can be thought about. Swimming, running and dancing are sets of body techniques that can be done without tools. They are, like all body techniques, marked by the cultural environment – dance more so than swimming or running because its objective is expression. They have this peculiarity that no one can mistake: they have to be learned and the *technical* difficulty is real. As much as we might think that walking, running or even climbing cannot be learnt, you just have to get started, whereas for swimming and dancing, everyone knows that you have to learn. Those who can't swim drown if they are thrown into the water, which was the classic case of sailors until the middle of the 20th Century, but not of the Ancient Greeks: Plato uses the expression "they can neither read nor swim" to describe the incapable or the uncultured (*Les lois*, Livre III), putting reading and swimming on the same level: an acquired know-how.

The use of a pole to cross rivers, to pass over bushes or hedges, or to leap while hunting is probably older than records show. While pole vaulting was not part of Greek sports and seems absent from their culture, it started as a sport in Germany at the end of the 18th Century. And since then, pole vaulting has been evolving! It has transitioned from the rigid pole made of ash, oak or cherry (probably descended from spears or masts) to the flexible pole made of bamboo (Olympic Games of 1900), aluminium-copper alloy (1950s), fiberglass or carbon fibre (1960s and beyond); at each stage, the previous records have been broken by a significant margin. With improvements in material, the body technique changes significantly; the current pole vault athlete must be a runner, a jumper and a gymnast. More generally, the revision of material technique requires a change in body technique, which was therefore possible. This ability to find the right body technique for a new tool does not seem to surprise anyone, as if it were obvious. Likewise, speaking calls for gesture, although if we are involved in a technical gesture, it slows down speech. If we talk about subjects that are important to us, it is rarely possible for us to do anything other than unconscious gestures. Mastering gestures and knowing how to speak can be learned and must be considered as corporal techniques; gesture and speech mobilize similar body capacities to the point of competition.

The mime is based on a body technique, whose origin goes back to a very distant past, one of the basic human characteristics; it involves the whole of the body; it requires knowing how to copy and stage situations and emotions. Mime shows a complexity similar to that of speech, although based more on signs than on symbols. As for dance, we know the incredible variety of possible expressions. In front of some dance performances, we can be troubled by a feeling that comes from what those who have been called for too long "primitives" call "the sacred". Dance can make us realize that the body is capable of expressing the sublime and the sacred. What is this specific technicality of dancer? Have those who are moved by dance asked themselves the question of the age of such a technique and the conditions of its existence? Is dance given to us when the tools are our inventions? Would the body have to be transformed to be able to be used as a tool or to be able to manipulate any tool?

We speak of the technical (or artistic) excellence of a pianist, by dissociating the quality of the instrument from the pianist's own know-how. Nobody, or almost nobody, asks the question of what makes it possible to acquire difficult technical know-how from a relatively new object. The same remark can be made in mathematics, of which we know today that without writing, this marvel of the human mind could never have appeared. There seems to be no technical object that can be said to require any particular bodily know-how to use it, even if it is only discovered in the face of the new situation. Everything happens as if this acquisition was self-evident. Gilbert Simondon in a television interview on mechanology in 1968 shows, among other things, that while there are many different hammers, for those who know how to use them, the use of a hammer is as much that of a tool as that of an instrument. It can even be used as a utensil. If the technical objects are the most visible part of the technique, they are not the only elements of it, and even, in some cases as we have just seen, they are absent[1].

1 This interview on Radio-Québec by Jean Le Moyne, "Gilbert Simondon Entretien sur la mécanologie" is on YouTube in three sequences: https://www.youtube.com/watch?v=VLkjI8U5PoQ; https://www.youtube.com/watch?v=HRqy9vttW-E&t=15s; https://www.youtube.com/watch?v=kCBWTHjKvbU&t=956s.

We cannot summarize the technique as the use of a tool, or even as a particular know-how, without taking into account the invention or the learning process. By way of a definition, we will propose here that *any technique, in its most general sense, is a know-how, which comes from original inventions, old or recent, which can be learned, improved and transmitted.* We can also talk about techniques, and we shall see later that there are indeed several categories of techniques. The advantage of first attempting a phylogenetic approach is that we approach the mode of existence of the technique, or techniques, from a dialogue between conception and confrontation with the facts. What is a human doing who learns a technique, uses it, teaches it? Little by little a different understanding of what a technique is is revealed, taking into account all its possible forms up to the present day. Phylogenesis and ontogenesis can then come together.

2.2. Awareness and use of techniques

According to this definition, *every animal has skills.* It is conceivable that it is a property of every living being. With birds and mammals, skills which are invented and/or transmitted and/or learned appear. It is recognized that *many species of birds and mammals learn from their parents, without whom they cannot become adults nor gain the skills necessary to lead the life of their species.* Much smaller numbers invent new behaviors and uses that are then learned and passed on. *Technical posture can be defined as an effort to invent, learn or transmit in order to acquire know-how in response to incentives, which more often than not come from the parents or congeners* (in the case of so-called social species), or from the environment – which includes congeners (fleeing, attacking, negotiating, feeding, reproducing). This definition covers both bodily and instrumental techniques, bearing in mind that the latter always require bodily technicality.

At this stage of our reflection, which is intended to be phylogenetic as well as ontogenetic, we do not question the essence of the technique. The essence of the technique is not technical, as Heidegger said. We will return to this subject at the conclusion of this study, which is also the construction of a narrative. Indeed, as we have just seen, focusing first on the question of the being of the technique poses a methodological problem. It is first necessary to situate the technical history in order to grasp its essence.

A technique is not always a conscious practice. By observation, we might even conclude that the use of many techniques is often unconscious; it seems to be the fate of a technique to escape consciousness as soon as it is mastered, to the point that mastery of a technique can be confused with instinct. For example, in chimpanzees, if a primiparous mother-to-be is isolated from her group and is in a human world, she will not be able to care for her newborn. She has to be taught that. We've only been able to find that out through this experience of isolation. Learning is not being conscious, at least in the sense of reflective consciousness. Even in humans, the first lessons were unconscious; consciousness gradually emerges, incomplete; even consciously acquired learning is quickly removed from the conscious field; in conscious learning itself, the whole process underlying the possibility of learning is not conscious. The construction of representations, which is necessary for some learning, does not lead to maintaining awareness of the know-how, which quickly acquires a form of unconscious automatism. Even in what seems to be the most conscious – the practice of a language – we are not aware of the way in which words emerge and follow one another. In rapid intentional action, we are unable to consciously think about the intention itself just before it is implemented. The different conscious representations of the same complex system cannot be maintained together; we move from one to the other without continuity. The example of the so-called rabbit-duck illusion, which has been circulating on the Internet, does not insist that it is not possible to keep both images in consciousness at the same time.

In short, we are never sure to learn, discover, transmit, in all consciousness. Technical practice and awareness of technique are independent. But above all, technical mastery and reflexive consciousness are opposed to each other in that they are only able to collaborate over time, alternatively and not in parallel. If consciousness came last, phylogenetically speaking, and technique is in the beginning, it is understandable that our consciousness of the role of technique is limited.

2.3. Technical posture in human phylogenesis

Since Leroi-Gourhan (1943, 1945), there has been a tendency to note that there is no human community without technique and that technique has played an important role in human evolution. Kenneth Oakley (1949), around the same time, recognized that the ability to make tools was a fundamental biological characteristic of humanity. As today's concern is more about the fate of humans within the current technological framework (Chabot and Hottois 2003; Benasayag 2010), it may be useful to rethink human evolution by considering technique, not as the late production of *Homo habilis* or *erectus*, or even *Homo sapiens, but as the technical posture that underpins hominid lineages.*

Like Peter Sloterdijk (2010), we will start from two pieces of information that we have now acquired: the technique constitutes part of being human, and humans are, in fact, neotenic compared to the great apes which are so closely related according to molecular analysis. Peter Sloterdijk took up ideas already put forward by Paul Alsberg in 1970 (Alsberg 2013). We will also use recently acquired data on the qualities of the closest great apes, whose lineages separated from those that eventually led to humans, about seven million years ago. Indeed, what we share with these closely related great apes, namely the chimpanzee and the bonobo, is likely to come from our closest common ancestor, even thought it was a population. I will take as a reference point these common qualities, which we now know to be more numerous than we assumed even 50 years ago. It must be recognized that human technique, in its complex, elaborate and systematic nature, distinguishes all hominid groups, whatever they may be, from the great apes, even the closest ones, which makes a comparison based on ethology stand out. It is a human characteristic. As has already been written, the technique is "anthropologically constitutive" (Havelange *et al.* 2002). Most philosophers, after Plato, often ignore or even despise technique: it is empirical. It is appropriate to return to phylogenetic analysis which, better than any speculative reasoning, will show us whether technique is a component of the possibility of thinking, and even more profoundly, a component of the broadening that defines possibilities.

Before considering the elaborate techniques that emerged in the late Paleolithic and early Neolithic period[2], and *a fortiori* those that followed, we will first focus on the technical evolution and its impact on hominization, that is up until the last "exit from Africa", which took place between 50,000 BP

---

2 Some anthropologists prefer to speak of Pleistocene (middle and late) and Holocene, following a geological approach. I will speak here of Paleolithic (archaic, lower, middle, upper) and Neolithic, separated by a fuzzy zone called Mesolithic, which corresponds to a prehistoric rather than a geological approach, although the coincidences are striking. The Neolithic corresponds to the beginning of agriculture.

and 70,000 BP[3]. Many researchers now recognize that technique and its evolution have played a very important role in hominization (Malafouris 2013); it is likely that its role is still well underestimated. Thus, it is necessary to carefully redefine how the technique and its influence in the process leading to the human being is established.

It is therefore a matter of using facts and data, and theories developed by paleontological researchers, as well as analyses by philosophers, anthropologists, geneticists, neurologists and cognitive scientists. We will simply put them in dialogue and build a set of coherent hypotheses, some of which shall eventually be validated, refuted or modified. The aim is to question empirical results and descriptions, researchers' hypotheses, philosophers' analyses and theories, and to confront hypotheses with the empirical data available in order to put forward a hypothesis.

In order to do so, we will propose a resolutely evolutionary, that is (neo)Darwinian, approach by analyzing certain crucial aspects which finalist approaches may not cover. The approach is empirical, yet without excluding ontological or even metaphysical reflection[4]. Two great thinkers have transformed thinking about humans by resolutely defining an empirical approach: Franz Brentano in psychology (1874) and Charles Darwin in biology (1859). Their filiation has been immense and includes, in fact, all current schools of thought in biology and human sciences. Today, this approach is no longer opposed to rational or experimental approaches; it complements them and the entanglement of these three research methods leads to the possibility of reconstructing a new ontogenetic and metaphysical way of thinking which would integrate phylogenetics.

---

3 The current convention is to note BP (Before Present, before 1950), for the Paleolithic and prehistoric dates, and for the historical or protohistoric dates we now note BCE (Before Common Era), and CE (Common Era) without Christian reference.

4 Empirical in the sense in which Franz Brentano speaks of psychology from the empirical point of view, that is neither rational nor experimental. Like Darwin, whose "empiricism" was reproached by Claude Bernard, Brentano has had a definite influence on phenomenology, psychoanalysis, new theories of knowledge, etc.

3

Ethology: Technique and the Frog

3.1. The Goliath frog: a technician frog

We all know that frogs are batrachians which experience two states during their lives: that of tadpoles, then that of adults whose jumps and swimming style remind us of humans. Frogs generally lay their eggs on plants and algae in ponds or along rivers, and tadpoles, which are vegetarian, feed on these plants and algae among which they swim. The mortality rate of tadpoles is high because they are preyed upon by many predators (fish, birds, reptiles, mammals). In fact, frogs generally lay their eggs wherever they can in the water or on the stems or leaves of plants that are immersed nearby. The adult frog, after its metamorphosis, becomes a carnivore and feeds on small, and more or less flying, insects.

This is in our temperate European countries, as we know of around 3,000 species of frog. Some of them try to gather their eggs in a sort of nest. But there is one region whose humid forests and river crossings are home to the largest frog in the world: the *Conraua goliath* or Goliath frog. It's hardly an ox, but it can measure 34 centimeters and weigh 3.3 kg. It is known over there, in Cameroon and Equatorial Guinea, and hunted for its flesh despite its vivacity. Most recently, researchers have lifted the aura of the unknown from their reproductive behavior. These frogs take care of their young by building beautifully designed pools along riverbanks (Schäfer *et al.* 2019).

The Goliath frogs living on the banks of the Mpoula River in western Cameroon build real pools, cleaned and consolidated by stone walls weighing up to two kilograms, at the edge of the shoreline, just at river level.

These real pools are one meter wide, sometimes protected by dikes. They even use excavated material to build the walls of these pools. These "nests" of Goliath frogs, at the river's edge, take into account the water level so as to have enough water not to be swept away by a variation in the river level and to be able to allow the tadpoles to leave towards the river by opening up a channel, at the right time. These basins could be used several times; the researchers found both tadpoles at the end of their larval cycle and eggs just laid. The males of the Goliath frogs would build the pond while their females would wait around. The male would call the female once the work had been completed, then the female would lay the newly fertilized eggs and watch them at night to guard off any predators. She would open the nest towards the river once the tadpoles had grown large. As many as fourteen operational basins were found in succession over four hundred metres, suggesting possibilities for inter-individual cooperation.

3.2. Causes for the Goliath frog's gigantic size

This construction effort, coupled with nighttime surveillance, combines an unusual degree of parental care and technical mastery in these frogs. And all the more so as the researchers found variability in the constructions and were able to define three types. The Goliath has invested in the technique, and this activity has become selective in its evolution. This is why, the researchers suggest, the Goliath has become so big! The larger the frog, the heavier and stronger the construction and the greater the protection for vulnerable tadpoles which have just hatched.

It is not known at this time whether Goliath frogs teach this skill to their young after their metamorphosis into frogs or whether it is an innate behavior like that of the spider whose web is an extension of itself. We also don't know if this will end up producing neotenic goliaths, with larvae producing protected eggs in their ponds.

In fact, what is certain is that questions remain: does the Goliath frog owe its evolution towards gigantism to selection of its technician skill? Does it teach its offspring this knowledge? Has its technique improved, is its brain more developed, or its legs differentiated? Another certainty is that batrachians are now joining the prototechnical groups of animals: primates and mammals (such as the beaver), birds and some reptiles. The practice of the technique is rooted in a very distant past.

4

Neoteny: From Concept to Grand Narrative

4.1. Sources of the concept of neoteny in biology

Could there be a link between the "neotenization" of the human being and the fact that he/she is a "technical animal"?

As soon as the word appeared, the concept of neoteny was marked by the seal of randomness, strangeness and myth. The story of its genesis begins towards the end of the 18th Century. The ambystoma (or amblystoma) is described as a rare blue-coloured salamander that could be found around the lakes of Xochimilco and Chalco and the volcanic craters known as axalpascos, at 2,000 meters above sea level, in the Federal District of Mexico; several species are described. At the same time, the axolotl, an animal well known to the Aztecs, who included it in their mythology, is described. The term axolotl means "monster from the water" (atl for water, and xolotl for monster); we only have to see a photograph of an axolotl to feel a disturbance linked to the strangeness and proximity of this apparently fragile, almost translucent animal, with gills floating around its head, and immense eyes that seem to be scanning the infinite. It rarely exceeds 35 centimeters; its monstrosity does not make it dangerous and is not without recalling the Heideggerian concept of a monster or that of Richard Goldschmidt (1940) who speculated in the phases of cladogenesis – that is to say upon forming a new group – on a "promising monster"[1].

---

1 Goldschmidt was a pioneer in research on the so-called homeotic genes that play an important role in the process of embryological development.

After its discovery, the axolotl was described by Europeans. It has the characteristics of a salamander larva. However, it can reproduce as it adapts. As sexual maturation processes at different stages of development were known in other phyla, scientists soon considered it a separate species; it belongs to the amphibian (or batrachian) class, a subclass of urodela, such as salamanders and newts, a class known for its remarkable regenerative abilities among the quadrupeds. Auguste Duméril, a French researcher in herpetology (the study of snakes), became, in 1851, assistant naturalist at the Natural History Museum of Paris; he created the vivarium, then brought axolotls from Mexico to reproduce there. The surprise came with the appearance of an adult salamander of the ambystoma type, which proved to be capable of reproducing, and whose young were axolotls that either metamorphosed into ambystoma, or, in the latter case, became adult axolotls. It soon became apparent that this was a single animal species with two forms capable of reproduction: one larval and one adult. Although a herpetologist, Duméril was interested in the conditions of axolotl metamorphosis. An article with his name came out just after his death in 1870: "Creation of a white race of axolotls in the reptile menagerie of the Natural History Museum and remarks on the transformation of these batrachians".

Various experiments were carried out by other researchers to understand the environmental or hormonal parameters under which the axolotl larva transforms into either an axolotl adult or an ambystoma adult. The axolotl became a laboratory animal, although its survival in Mexico is now questionable. It seems that the term "neotenia" – from the Greek νέος (new) and τείνω (expand) – was proposed by the Swiss biologist Julius Kollmann in articles published in 1884 and 1885.

4.2. Applying the concept of neoteny to the human being

The concept of neoteny was first developed based on the theory of Louis Bolk (1926), a Dutch biologist, known as "the theory of fœtalization", according to which man would preserve the fetal characteristics of his related species (bonobo, chimpanzee). It must be acknowledged that the bonobo baby, at birth, has proportions that resemble a three-year-old human child and that the head of the bonobo fetus has a human profile. His text was translated into French in an issue of the French journal of psychoanalysis, 35 years later, showing the more general cultural impact that these studies were beginning to have on the concept of maintaining infantile characters as far as humans are concerned. Even though Bolk was strongly criticized in the world of biological research for his simplified and outrageous generalization, the relationship between his thesis and the concept of neoteny will be made, all the more so since at the same time, Émile Devaux (1921), remaining partly external to academic activity, focused on the comparison of the development of the great apes and that of man and showed the relative "infantilism" of the latter. Would humans be to great apes what axolotl is to salamanders? A child capable of reproducing?

Today, we use the term paedomorphosis in two senses. On the one hand, we use it to mean the progenesis which consists of accelerated development resulting in earlier sexual maturity and the formation of small adults. The best known example is *Paedophryne dekot*, the smallest known frog (another batrachian!). On the other hand, neoteny is the opposite process, in that it is a deceleration of development: the adults are a normal size while retaining juvenile characteristics. The animal most often cited as an example for neoteny is the axolotl, as we have seen. In particular, it has gills and not lungs, which means that it cannot get out of the water, whereas the "normal" non-neotenic adult has lungs like any adult newt or salamander and can go on land. As far as the Goliath frog is concerned, it's anyone's guess.

Peramorphosis, which describes the opposite phenomenon (i.e. either an acceleration of development or the acquisition of hyperadult characters) is of interest for our purposes only because this phenomenon is the result of common processes called heterochrony, which involves changes in the temporal rhythm of development affecting certain organs. Neoteny is a particular case of phenomena now identified and studied in the development process of an individual from fertilization to the adult state. This process is called ontogeny.

It has come to be understood that this is the name of related biological phenomena grouped together in the expression "heterochronia of the development". Organs or groups of organs develop at a different rate from the reference standard, which corresponds to the species from which the individual originates. Molecular understanding of these phenomena is beginning to become quite advanced. A single mutation in an important regulatory gene during ontogenesis, or a change in environmental conditions affecting the expression of this regulatory gene, can cause such heterochronies.

In 1960, Franck Bourdier, a French prehistorian, used the concept of neoteny to explain human evolution. In 1986, Peter-Andreas Gloor, a Swiss anthropologist, published a text recalling the history of the concept of neoteny by linking it to that of the Oedipus complex of psychoanalysis. He shows that the late onset of biological sexuality is a factor of hominization, since it is absent from the primates closest to humans and is amplified from *Homo habilis* to *Homo sapiens*. The relative delay (heterochrony) of sexual maturity thus allows a prolonged childhood, in other words a relative neoteny for the development of the brain.

Stephen Jay Gould, an American biologist, took up Block's work again in 1977 and showed that, although it is not possible to take up all of his concepts, they can be accepted by recognizing differentiated heterochronies, that is differentiated modifications of the rhythms of development or growth, during ontogenesis, in particular in the respective shapes and sizes of the parts. These heterochronies, in humans, may not be neotenic, but peramorphoses, such as the evolution of the pelvis (Berge and Ponge 1983). Moreover, the lengthening of the legs is a differentiated heterochrony, without being neotenical; it is not a childhood characteristic. These are the concepts that will be taken up by paleontologists and developmental biology researchers to evaluate the different stages of hominization from *Australopithecus* to *Homo sapiens*. The multiple heterochronies, resulting from specific mutations, could be an important key to the evolution which led to humans.

4.3. Appropriation of the concept of neo-neoteny by the humanities

Desmond Morris, a British zoologist, takes up the idea in *The Naked Ape* (1967). Human biological infantilism, known as neotenicism, is beginning to spread outside biology. It appears that two complementary theses – even if they do not present themselves as such and sumptuously ignore each other – are developing: man as an equipped living being (from Ernst Kapp, to Marcel Mauss, André Leroi-Gourhan, André-Georges Haudricourt, Gilbert Simondon, Lambros Malafouris, Krist Vaesen, etc.); and man as a neotenic primate (Konrad Lorenz, Arnold Gehlen, Gavin R. de Beer, John B.S. Haldane, Desmond Morris, Ashley-Montagu, Stephen J. Gould, Edgar Morin, Delbert Thiessen, Dany-Robert Dufour, etc.).

Gilbert Simondon uses the term neotenization to speak of human individualization by dedifferentiation and reincorporation of pre-individual reality into a new individualization (Morizot 2011). This notion, scientifically founded and taken up according to a philosophical approach, could be fruitful in understanding what has been happening for five centuries. But it is the thesis of Dany-Robert Dufour (1999, 2012) that will be of interest here.

Through an admirable use of metaphor, by giving some of his texts the appearance of fiction, Dany-Robert Dufour wants to convey the idea of the lack, incapacity, fragility, even error of nature, of this being that is human. For him, the human, an obvious neotene, is an unfinished primate, just as the axolotl is an unfinished, maladjusted ambystoma, incapable of surviving on its own; this imagination goes beyond the axolotl which, in its confined, stable and protective environment, is adapted. This leads to the idea of a second nature.

Dany-Robert Dufour refers to Anaximander of Miletus's remark 26 centuries ago: "In the beginning humans were born from animals of different kind, since other animals quickly manage on their own and humans alone require lengthy nursing. For this reason they would not have survived if they have been like this at the beginning" (McKirahan 2010, p. 42).

Anaximander was also the first to assume a marine matrix to man. He demonstrated a first approach that we would qualify today as phylogenetic. Without this prior second nature, how could the first, the one that appears to us today, be developed?

Dany-Robert Dufour undertakes an ontogenetical analysis; he posits that Man has "invented" his second nature, rather than that the latter has allowed him to acquire, in the long term, this "original nature". Doesn't the ontogenetic analysis need to be completed by a phylogenetic analysis? We know today that ontogenesis does not summarize phylogenesis, even if traces of the latter remain in the former. It is these traces that have made it possible to set up research in evolutionary developmental biology known as "evo-devo" (Chaline 2006b, 2014). This work has linked phylogenetic data and the understanding of the fine molecular mechanisms of embryogenesis with the discovery of homeotic genes, which code for the initiation of successive stages of development.

It appears that, since the 1970s, the concept of neoteny has left biology; it is used by anthropologists, psycho-sociologists, psychoanalysts, philosophers; it invades fiction: literature, cinema, animated drawings; it has been transformed, in biology, from concepts such as developmental heterochrony and exaptation[2] and as a result of the evoked studies mentioned above.

Neurology renews, by confirming it, the neotenic hypothesis. In doing so, it clarifies the terms according to the evo-devo approach, which combines the theory of evolution (phylogenesis) and theories of development (ontogenesis). Alain Prochiantz (2001, 2010) concludes that man is anatural, that is, he has somehow emerged from natural processes, although he comes from it; he does not go so far as to say that he is artificial. Jean Chaline (2014) arrives at a renewed proposal of hominization according to the same logic. Combining neurology, cognitive analysis and biology, Robert Bednarik (2012) shows that the neotenization of the homininae necessarily began millions of years ago; already, in many aspects, *Australopithecus* show neotenic aspects in relation to the other great apes.

4.4. Neotenization: a "grand narrative" of the emergence of the human being

Neoteny, infantilization, new capacities outside nature; this neoteny has a perfume of inexplicability. How could neotenization, which would go on to lead to the modern, inventive, creative and playful human being, have been stabilized before the invention of the tools that would give it the meaning it has today? Indeed, since the thinkers of antiquity, we have known that man, without tools and without an important society, cannot survive: he has neither claws nor fangs, and yet he is the first predator.

Marc Levivier (2011) analyzes neoteny, according to what emerges from the works of recent thinkers – Georges Lapassade, Giorgio Agamden, François Châtelet, Joseph Gabel and Dany-Robert Dufour – as an underlying, Western "grand narrative". The term "grand narrative" refers to Jean-François Lyotard (1979). For Franck Tinland (1977), based on the work of Claude Lévi-Strauss, anthropological specificity is described as a

2 Exaptation is a functional movement of a structure by selection. An organ can change function or participate in a new function; for example a thermal control organ that becomes a wing.

correlate of the withdrawal of natural characteristics and the development of artifice (language, tools, etc.): "The open gap in this living being between the demands of life and the natural dispositions capable of responding to them is undoubtedly the effect of the artefacts themselves." Peter Sloterdijk (2010) expands on this analysis and defines four so-called ontological-anthropological conditions, including neoteny and protection by artefacts, to transform a great ape into a man. It appears consistent; it takes both neoteny/heterochrony and protection by tools to "produce" a human. That's what must have happened, but how?

What underlies the reflections on humanization through neoteny is, on the one hand, the affirmation of incompleteness, and on the other hand, the reference to childhood, which is – it is an undeniable fact – a time of learning in all mammals and even more so in higher apes. Neoteny is also shown by cerebral plasticity. Neoteny would give humans this increased ability to learn and invent, even in adulthood, which again seems to be a human characteristic. Remaining a child, humans retain the capacity for learning and discovery for a long time. We know today that, although childhood is a particularly long time in humans, sexual maturity comes years before the brain is fully developed. Reproductive ability comes late, but the period for brain maturation is even longer. We could speak of a double neoteny, sexual and then cerebral. Young couples who give birth to children are still in the process of learning and undergoing brain development.

These analyses almost always lack the precise relationship to the technique used, to the various known or supposed artefacts associated with the neotenization at the origin of humanity; if this neotenization tells us anything, it is certainly not that we had to invent a second nature; on the contrary, something else, which is protective, was needed so that this "neotenical nature" could flourish. We have seen, with the example of the axolotl, that under conditions of stabilized and slightly high temperature, sexual maturity occurs before the adult transformation.

As all thinkers since Anaximander have noted, man/woman cannot survive as he/she is in a natural environment; socialization, culture and technique are the ingredients that he/she needs to become himself/herself, that he/she needs in his/her individual and personal development, that he/she needs to survive.

We can posit that this neoteny would be more a consequence in the phylogenetic process of hominization than a cause, which does not prevent it, step by step, from being a condition for the possibility of creating new tools and inventive capacity. Socialization, culture and technique would therefore have appeared before the process of neotenization that would be the consequence, while considering positive feedback loops, at each additional stage of neotenization, with socialization, culture and technique. Each socio-technical stage would lead to a selection that would favor a new stage of neotenization. This interaction between sociotechnical culture and biological and neurological development is well illustrated by the Royaumont Colloquium on the Unity of Man (Morin and Piattelli-Pelmarini 1978). It is understandable that there has been co-evolution, without further fine-tuning. This does not change much in ontogenetic analysis, but assuming that it allows us to understand the phylogenetic process, it would open the door to analyses and research integrating technical evolution and human development. This could help us to understand the new interaction between technique and humanity that has been taking place since the industrial revolution and to assess both the impact and the meaning it has for us today.

How exactly can technique support a neotenization? Is there a defining order? Is it determinism or openness to what is possible? And which techniques are for which neotenizations (and vice versa)?

The proposal is as follows: neoteny was impossible before technique and social organization were in place. Technique precedes neoteny; the latter began well over two million years ago. So, what about technique?

5

Issues of Neoteny and Technique

5.1. A very old conception of human "disabilities"

It is striking to note that since Anaximander, the first theorist of evolution, no one has continued to reflect on the fact that those who began the line from which we come cannot have been as we are: beings which are certainly unfinished, lacking, devoid of the capacity to directly influence nature and require a significant period of dependence to acquire their autonomy as adults, which they only obtain through the mastery and permanent use of many tools. Pliny the Elder (2015, Book VII) six centuries after Anaximander, was astonished that this master of the world – which he already considered obvious – began life as a naked and crying baby. How is it that a being who dominated others starts off as a crying baby?

From the oldest myths to the research of pre-modern philosophers, while natural human incapacity was recognized, it was assumed that technique was brought by the gods, or their acolytes, to humans once they had fully become human, which was paradoxical. Or that it was created by naked and helpless humans, somehow capable of recreating everything by themselves in order to survive (Adam and Eve driven out of Paradise). Neotenic beings: sheepish, whinging superheroes! So it is in Greek mythology; in order to allow men to live, Prometheus brings them fire. As if the ontogenetic analysis of humans led to the idea of the construction – by them or thanks to the gods – of a protective environment, and not to the idea that this ontogenesis, phylogenetically speaking, was possible by a prior protective environment, brought about by technique. The axolotl itself remains in adulthood under conditions of temperature and stable light; it is the particular conditions in a protective environment that induce this heterochrony. The ancient Greeks knew that man is an "animal with tools", but they interpreted it according to the logic of the myth of Prometheus. The Hebrews don't have any musings on this subject. Imagining an evolution which led to the human being was impossible (except for Anaximander, pioneer of evolutionary thought twenty-five centuries ahead of his time).

5.2. The equipped human and neotenic human: two unrelated concepts

Those who insist on the definition of the human as a being with tools (Mauss, Oakley, Leroi-Gourhan, Haudricourt, for example) or as a cultural being (Sahlins 1980; Sahlins 2009; Godelier 2010; Testart 2012; Stringer 2012; Vaesen 2012; Vaesen 2013) insist little, if at all, on the unfinished, neotenic aspect of the human. They often ignore the problem. This aspect, however, transpires from the description of the multiplicity of human societies, which leads to Simondon's concept of neotenization. In fact, we are interested in tools, culture, ethnology, ethology, sociology and biology, albeit separately. In order to search, you need a line of research, and it's easier if you stay within the same discipline.

Gilbert Simondon, followed by Bernard Stiegler, whose objects of study are recent techniques (metal tools, machines, digitization), barely took the oldest possible relationship between technique and human beings as a field of analysis before the emergence of the industrial world, and even less so before the historical period. Gilbert Simondon (2017) shows that the technical object serves as a mediator between humans and nature, and he concludes that this medium can, by its very existence, lead to the withdrawal of humans from the natural world. As his approach is ontogenetic, he does not think that this phylogenetic retreat can have been achieved; how did it establish itself? His area of focus is more the current evolution of the human-technical relationship than its relationship in the evolutionary period; his analysis of individuation is ontological. However, it may be judicious to think about this relationship by noting that such a transformation has already taken place several times, in several stages, over a long period of time, during the phylogenesis of the lineage of homininae leading to humans.

Most evolutionary biologists attempting to think about their discipline are moving away from substantialism and, in doing so, are moving away from classical Western philosophy. Jacques Monod (1997) in the title of his book refers to Abderitian physics; he begins with a quotation attributed to Democritus. This philosophy, from which Plato and Aristotle deviated, was related to Epicureanism. Considered materialistic, the physics of the Ionians and Epicureans was rejected by the majority of philosophers. However, it has found a beginning of consecration in the revival of the concept of the atom by researchers in chemistry, and more so by the Darwinian thesis, even if Darwin does not quote those thinkers who could have influenced him. However, they are precursors of a modern thought: that of evolution.

There was, from its inception, an apparent ontological weakness in Darwinian thought, because it was focused on the temporal process of evolution. You can see that it's evolving, though you can't see what's evolving. Thinking about being and thinking about evolution seems antinomic, because it is not the beings that evolve, it is the succession of beings that reveals evolution. It is not an evolution of the being, but an evolution in the being – which then transforms what the being is said to be – or an evolution of the beings in their temporal succession. It can be considered that this has prevented the involvement of philosophical thought, except, it seems, in Nietzsche (2011), through ambiguities that are sources of ideological and political abuse, and in Bergson, at the price of an orthogenetic approach (vitalism) that can no longer be sustained[1]. Times are changing, since Gilbert Simondon, through his criticism of hylomorphism and his apology of the Ionian pre-Socratics, has opened the door to a philosophical reflection on the Darwinian approach, as Victor Petit and Baptiste Morisot[2] clearly show.

Gilbert Simondon (2013) emphasizes the influence of technique in ancient Ionian thought; he describes Ionian thinkers as technicians, while Descartes is a thinker marked by the technical model (e.g. the concept of sequencing). However, he does not think of technique as a means of transformation, as a condition for the emergence, or even as a catalyst, of the human being. Even if he knows Darwin, he does not grasp the complementarity that some of his readers guess. Simondon's process of individuation requires a stochastic process, just as his description of capacity for transduction incorporates an underlying probabilistic selectionist approach. As Edelman and Tononi (2000) note (author's translation): "One

1 *Orthogenesis*, a theory now abandoned, states that an evolutionary lineage goes in a predefined direction, even if the duration of evolution lasts millions of years.

2 See Barthélémy, J.-H. (ed.) (2009, 2010, 2012). *Cahiers Simondon*, 1, 2 and 4.

can speculate that there are only two modes of structuring thought: selectionism and logic. If a third could be discovered or demonstrated, it would be an event in the history of philosophy." Transduction as a structuring way of thinking can be thought of, by digression, as selectionism.

Darwinian thought, accepted in a century by the majority of biologists seeking to conceive their discipline, has scarcely integrated the use of tools in its reflection on human evolution, and this despite the fact, which is fairly quickly accepted, that the species prior to, or ancestors of, *Homo sapiens* already had tools[3]. It remains confined to a biological approach, and the so-called Darwinian sociobiological attempts have not been convincing[4]. In fact, one element is missing. Even if many paleontologists integrate the impact of the technique into the evolutionary process from Australopithecus to *Homo sapiens*, few theorists, it seems, attempt to link biological, anthropological and paleontological data through more cross-disciplinary analyses.

5.3. The philosophy of technique[5]: a recent discipline

What prevents us from seeing the problems of this interaction between technique and humans over time on the one hand and its probable morphological and biological consequences on the other? We know that philosophy and science emerged from technique long after it, and not with or before it. Simondon shows that the first Ionian thinkers were marked by the technical approach. There was, in fact, despite a long break, a consonance between philosophical thought and scientific thought, made visible with the emergence of modern scientific thought from the 16th Century onwards. Technique, necessary for this methodological thinking, remained unthought of, even unthinkable, and yet necessary for scientific activity. It could not be thought of as a component; it is constituted or offered, because one of the characteristics of technique is precisely that it integrates with the human being to the point of going unnoticed, as Xavier Guchet (2005) has noted.

3 In *The Descent of Man*, although Darwin understood the proximity of man to the great apes and that the conditions of existence exert a selection pressure, he does not mention the use of tools as a condition of selection.

4 Spencer, in *The Principles of Biology*, 1863, and *The Principles of Psychology*, 1855, shows that he is more Lamarckian than Darwinian.

5 The term technique is used here as it is a reference to the Greek techne, see https://en.wikipedia.org/wiki/Techne.

The Greek technical system was very elaborate, although Greek philosophers make little mention of it, and it is forgotten that they could have had access to the steam engine and to industrialization if they had had in their technical system the ability to make alloys of materials resistant to high internal pressures such as, for example, steel. That social organization and cultural and intellectual foundations may have been lacking is possible, but we will not be able to establish that. Nonetheless, technically speaking, one necessary aspect was missing, as Simondon pointed out: materials for pipes and vessels resistant to internal pressures of several bars, which did not appear until more than 1,000 years after the end of the evolution of the Greek technical system.

There is thus a general underestimation of the importance of the role of technique through philosophical thought. During the 20th Century, it was gradually recognized as important, albeit often negatively, while the apology of progress was being reduced without disappearing. Is it not because its increasingly rapid evolution has made it tangible? It can no longer go past without being seen; it has become so important that it takes between 15 and 20 years to acquire the indispensable tools to live in today's society in the so-called advanced countries and sometimes nearly 30 years for the most complex professions. We can see, in our so-called "postmodern" era, that technical evolution leads to human evolution. The question of the integration of digital media, computers, smartphones, robots, biotechnologies and nanotechnologies is becoming obsessive (Ellul 1954; Stiegler 2013; Jarrige 2014; Tisseron 2015). Is this not an indication that we should go back to technique and analyze its role in the emergence of humans?

Although technical models were used in scientific and philosophical thinking, they were not thought of as such (Simondon 2013). It was only with the thinkers of the Renaissance that technique began to become an object of interest, more through an exchange with art than with philosophy (see Leonardo da Vinci, Bernard Palissy, etc.). With the Enlightenment and the *Encyclopédie* by Diderot, d'Alembert and de Jaucourt, the technique is given to be seen in its finest details, while Benjamin Franklin asserted that "man is a tool-making animal". The relationship described is that of the human being producing tools, and what about human-producing tools?

Charles Darwin's *On the Origin of Species,* published in 1859, was soon translated into all European languages, while Ernst Kapp's book *Elements of a Philosophy of Technology,* published in 1977, was only translated into

English in 1978 and into French in 2007. Darwin's work has a huge posterity; Ernst Kapp's has had difficulties. The causes are identifiable: his theory of the unconscious organic projection of the subject creating a tool is not very credible; his book combines disparate information, from which a coevolutionary vision between the human and technique does not emerge; and although published after Darwin's work, it does not follow the Darwinian model and is still Hegelian. Ernst Kapp's thought would nevertheless be taken up, in part, by a few thinkers: Ernst Cassirer, Alfred Espinas, Georges Canguilhem, Raymond Ruyer. Indeed, this theory, revisited by a thought stemming from phenomenology, epistemology and Bergsonism, suggests more than what it says verbatim; for example, the idea that, in order to know himself, man must make a detour through technique and his objects; or that in invention, the inventor is rarely aware of the technical models used; or that a technique is rarely developed first for the use that will prove to be the most important. Reading Kapp in the light of Nietzsche, Mauss, Leroi-Gourhan, Simondon, one can see the hidden gem among the considerations that today seem of little interest.

Consequently, a reflection on the relationship between the neotenizations of homininae and their technical development has been slowed down by: the lack of interest in technique on the part of philosophers; the lack of interest on the part of evolutionary biologists in the impact of technical use on human evolution; the difficulty for philosophers to think about biological evolution and the technical evolution of homininae, even separately.

Nevertheless, the work of Leroi-Gourhan, Franck Tinland and Peter Sloterdijk, and later François Sigaud, paved the way. For them, artefacts have played a role in human biopsychosocial evolution. Neotenization stems from this. It's thus a question of grasping the processes involved.

6

Neoteny and Fetal Consciousness

6.1. Humans before birth

Human neoteny, from the perspective of brain development, is the existence of a fetus with a large brain, in which neuronal connections are set up late in life in relation to the environment and human surroundings, and which is not mature enough for an independent life just after birth. This brain continues to grow, intra-utero, before differentiating. The infant needs several years of care, protection and relationships before beginning to acquire autonomy. Its huge brain is operational from birth, especially to learn continuously and quickly from the human world. There is surely consciousness even before birth, but what kind? Several approaches to the state of consciousness of the fetus, that is during the last two months of its development, have been put in place, more or less academically:

– neurological and philosophical approach: how does the fetal brain evolve in the last months before birth?

– clinical approach to birth: can the state of consciousness of a neotenic (human) baby be captured at birth?

– experimental approaches to the state of regression or de-individuation: can we relive a state of consciousness that is dedifferentiated, natal or prenatal?

For Francis Kaplan (2008), it is a question of grasping the difference between “being alive”, which applies to any organ in a living being, and “being a living being”. In the process of ontological individuation, at what point does the status of an embryo change to that of a fetus and acquire an individuation? How does the embryo leave the status of an organ for the status of a fully-fledged individual with psychic autonomy potential? It is when neuronal connections are established and autonomous neuronal activity is detectable that the status of a living being is attained and that potential individual viability is recognized. It is at this stage that it acquires the quality of a fetus. A developed brain coincides with a process of individuation that leads to psyche, interaction and the construction of otherness. Human individuation may be constructed even before birth, although it is still early compared to that of higher primates and will require a longer period of maturation in childhood. This is difficult to deny, because the infant learns the human world more quickly and more deeply than small bonobos or chimpanzees immersed from birth in a human world. Compared to species of the *Pan* genus, the human brain at birth has grown, especially from less specialized, modular and strongly interacting brain areas, which can develop new neuronal interactions, absorb information from the senses, process and reprocess it according to the principle, analyzed by neurologists, of functional coordination and integration through re-entering loops. A state of consciousness is a unified state (Edelman and Tononi 2000).

According to Jean-Marie Delassus (2001, 2005), birth is not childbirth. It is for ease of thought that we believe in a natural framework that leads in a linear fashion from the development of the embryo and the fetus to its birth, beyond the birth of the child. For Delassus, this is a mistake. Human birth invites us to discover an unprecedented dimension of existence, because, if everything is ready for birth, the experience of the baby being born goes beyond the event of childbirth. This is in line with the concept of "trauma of birth" by Otto Rank (2015). Even if this concept now seems out of step with the meaning of this event – the process of passing from the fetal to the infant state – it has had the advantage of making people aware that this is a very special moment. The image of the screaming, impotent newborn baby in fact masks a sovereignty: that of life in its still original state. No parent is mistaken and, except in specific circumstances or contrary lethal cultural pressure, everything is done to respond to this need to live.

Through the newborn, we get a glimpse of who we are. A baby, powerless and defenseless, who demands that everything revolves around it, whether that be food, temperature control or cleanliness. Psychoanalysts call this omnipotence. To envisage this core of the unknown hidden at the heart of what the word of birth cannot name, Delassus elaborates a "psychoanalysis of birth". According to him, the rigorous clinical observation of the newborn child, put in relation to the prenatal life that structured it, makes it possible to carry out this attempt. His two founding works propose a complete path from prenatal, fetal life to the completion of early childhood. It details the constituent stages of psychic birth. The fruit of exceptional clinical experience, Delassus' works invite us to rethink the multiple dimensions of birth in humans. Birth is also co-birth[1].

6.2. Humanity of the baby at birth

The human baby possesses all the capacities to integrate what will bring it into the human community (Mehler and Dupoux 1990). It is neotenic, since on the one hand it is born late, but also too early, and thanks to its bulky brain, certain developments, even before birth, are properly human and allow it to make use of the time available after birth. Its development will be built in and via multifunctional interaction with a techno-human environment that it is able to integrate. This is what we call the four E's of cognition (extended, embodied, embedded, enacted). It is apparently born too early, after a long prenatal development that allows it to have a brain with infinite possible neuronal connections, be capable of learning continuously and for a long time, and continue to build itself through its interactions with its environment. Contact with its environment will be a stimulus for further brain development.

According to the third approach, which could be called "experiential", or clinical, therapists and researchers have proposed carrying out specific experiments aimed at simulating the crucial experience of birth, made of the encounter with the unknown. Leonard Orr (1977) and Stanislav Grof (2010), independently, developed a breathing technique that would lead to an experience identified as rebirth, which some believe to be the experience of birth. Even if there may be a bias in these individual experiences, because it is not a question of reliving the fetal and natal experience in a raw form but of accessing memorized and more or less reconstructed traces, the fact remains that the narratives testify to an experience of undifferentiated consciousness and, as Delassus suggests, an experience of totality. The discourses are moreover surprising because they recall, without the authors being aware of it, the conception of being according to Parmenides and then

1 This is a play on words which works successfully in Delassus' native tongue in French, *connaissance* (knowledge) and *naissance* (birth). We have decided to use the term co-birth.

Plato: one, total, encompassing, undifferentiated, spherical. We also find these semantics in Delassus.

It is highly likely that the neoteny of the fetus, which has a huge brain compared to that of the large primates closest to humans, will give it a form of (pre)consciousness, an ability to become fully human, provided that development continues in a protective techno-human environment. To be able to survive, human beings at birth must benefit from a minimum protective environment, which can be described as socio-technical, even if it is most often described as solely relational and affective, whereas this relationship is not specifically human, as the bonobo or chimpanzee baby also requires it. In any culture, child protection is always an elaborate technical approach, even if this technique favors, or at least does not contradict, direct contact with the mother. The few known attempts in history to voluntarily separate newborns from any human presence have all ended badly. But Elisabeth Badinter (1982) described a dramatic situation in the 18th Century, showing that social reasons can also lead to the opposite of what the technique would allow. The instinct that the technique supported proved to be fragile, neotenic, because the mother herself is neotenic.

6.3. Ancient protection of the human baby at birth

This understanding of the "baby state", specifically the human one, seems important to reflect on the phylogenesis that led to humans. It should be integrated that these characteristics of the human baby were the same in 170,000 BP, and *a fortiori* 50,000 years ago. That is to say that the "prehistoric men" who went to Australia, those who crossed the glaciations and walked to Alaska and then all over America, were born like us, and had the same needs during their first years. How was it possible? How could humans at those times adequately protect and provide a suitable environment for infants in a hostile and dangerous, new, often icy, sometimes hot, environment with which they were, it seems, almost in direct contact? Were they not more "armed" than we imagined?

The answer might be that even though life expectancy was not so long, infant mortality and maternal mortality were high, while human groups from those distant eras had sufficient child protection tools and capacity to live in harsh environments and ... increase their populations. It is likely that climate-related selection has remained strong. This can be seen in certain body aspects of different populations that are "adaptations" to a lack of light or cold (decreased skin pigmentation, length of limbs, shape of nostrils and size of nose, slanting eyes). Nevertheless, surviving such strong and rapid changes in climate means that technical protection has been very effective.

7

Inversion of the Analysis: The Lamarckian Bias

7.1. The ambiguous concept of adaptation

Reflecting on the processes of hominization is a difficult exercise: many paleontological archives are still missing; the researcher is impaired, as it concerns the issue of the human in himself; we are all marked by a teleological approach that corresponds to an unconscious conception of our way of being. It concerns our emergence. It concerns the appearance of a new state of being, but how and from what?

In the 18th and 19th Centuries, the problem was similar for transformative approaches. Darwin recognized the lack of archeological and geological data, which made it easy for criticisms to be refuted from then on. In the century that followed, his approach was dramatically confirmed by numerous fossil discoveries, not to mention other approaches (population genetics, molecular biology, developmental biology, etc.). Here we can, like Darwin, accept the "lack of archaeological data", and put forward hypotheses that can be confirmed or refuted in the long run by studies and observations, as well as by molecular studies when possible.

The bias qualified here as Lamarckian consists of wanting to think about the adaptation of homininae lines which thus acquire new capacities, because individuals adapt, whereas it is a question of understanding the process which made it possible to select in various successive stages, over approximately seven million years, the pre-human then human populations which appeared between 300,000 and 70,000 years ago. Many authors show that cumulative cultural evolution in humans causes the rapid evolution of human populations compared to that of today's chimpanzees or bonobos (Dean *et al.* 2012). The question posed here is primarily about the process that led to this difference. It's not a question of how humans evolve today, but of how higher primates without an initial accumulative culture gradually acquired this ability, to the point where humans with a taste for accumulation have an almost compulsive side. These are the beginnings of wealth...

There is a specific difficulty in thinking about adaptation. The term adaptation is used in at least three different senses, and to be Lamarckian is to believe that these three senses are similar. Individual adaptation is a quality of the living individual that has within it the ability to cope with change; certainly primates, and much more so humans, possess this ability at a high level. This same adaptation can concern a more or less large group. The adaptation of a population over time describes the appearance, selection and stabilization of a new character over many generations by natural selection. Carriers eventually pass on the trait in subsequent generations, which spreads through the reproductive, biological and/or social process. For a species that lives in communities, this novel trait does not always need to be general. It may be sufficient for a certain proportion of the population to express it. This limited and sufficient aspect of the spread of traits within a population does, in fact, create intra-population variability. This situation gives rise to a third type of adaptation which describes the plasticity of a population at a given point in time, which will be all the higher as individuals carry many different characteristics distributed throughout the population and as they themselves become adaptable in the first sense. In a discourse on adaptation, the three different senses become intertwined, which can lead to misrepresentations.

Moreover, the transmission of adaptation involves several different levels as soon as individuals of a species can, through learning, acquire new behavior. Anything in the learning order, including, in principle, techniques, can be transmitted and also lost; this transmission is done at an individual level, but can have a collective impact. This behavioral or technical variation resulting from learning – which is a transmission – can be selected and therefore adapted in the population sense. The same term, "adaptation", is thus used in different senses, which depend both on the real capacities of individuals and on the level of scale considered. Care should be taken not to confuse them. During a plague epidemic, as in the 14th Century, adaptation was the surviving population.

This shift in point of view leads us to reflect on the order of phylogenetic successions. A Lamarckian approach would be to say that the appearance of big-headed babies led to the enlargement of the pelvis of hominid females, that is unlikely double selections. The Darwinian inversion would be to say that it is the enlargement of the pelvis, previously, for other reasons, that led to the birth of larger-headed babies – a trend that seems to be fairly general in mammalian lineages[1]. Likewise, it would not be the insufficiency of this enlargement that led to the birth of premature babies (compared to the developmental stage of the superior small primates: chimpanzee, bonobo); premature birth was favored or allowed by specific, more efficient, protective conditions at birth, which may have occurred previously, for other reasons related to technical evolution, which allowed both relative brain growth and selected heterochronies of the neotenic type in the birth process. Similarly, mastering a technique physically does not mean that one has a prior representation of it; it may appear long afterwards. The action preceded the representation, just as the intentional representation may have preceded the verbalization or visualization of that representation. Having a representation of an object, in association with an action, does not mean that awareness of the representation, that is reflexivity, is present. No tennis player has reflexivity about his movements during a match.

7.2. The uselessness of adaptation with the concept of natural selection

When a new collective characteristic appears, it has been chosen; selection conditions must be present before the characteristic emerges. This does not mean that a neutral mutation cannot appear and persist prior to the selection that will allow it to spread throughout the population. The establishment of a new trait in a population requires prior selection pressure favoring the identified trait over a population that is heterogeneous with respect to that trait. For example, it is now known that bonobos have an amazing ability to learn a pre-language of signs that allows them to express

1 Recent discoveries about Huntington's disease suggest that this human degenerative disease comes from the extreme multiplication of a codon of the HD gene. However, this duplication starts at the beginning of chordates, with fish, amplifies in mammals, even more so in apes and even more so in humans to the point of reaching a critical threshold that could be shifted by another mutation elsewhere. Each time this duplication increases the amount of gray matter (thus neurons) of the species under consideration (Lo Sardo *et al.* 2012).

their needs. In their natural context, they do not need this ability. However, it is this predisposition that was selected during the emergence of language towards *Homo*. Certainly, his ancestors did not live in the same context. In the same way, a technician species can lose a technique, for lack of use, and find it again a few hundred or thousands of generations later.

We know today that it is not from the formulation of a law or a rule that a tool, instrument or utensil is created; on the contrary, it is from a tool, instrument or utensil that it may be possible, often long afterwards, to recognize and then formulate the corresponding law. Ernst Kapp had already understood this. Every throwing weapon, every striking object obeys the calculation of kinetic energy; yet the calculation came hundreds of millennia after its use. Ships have been built for tens of millennia without understanding the laws of buoyancy or the effect of the wind on a sail; the use of oars, paddles and sculling was as well prior to the understanding of how they worked. The invention of the motorboat propeller is recent, and the theory to understand its operation did not exist before its design, which was the result of events not foreseen by the inventor; the same was true for the steam engine or the recent use of composite materials, etc.

It is therefore important to distinguish, on the one hand, the process that leads to a new product, use, or event, and on the other hand, the meaning (or calculation) and its displacement that emerges, most often after the fact. There is generally no possibility of deducing the latter from the former. This is because it has been possible, recently, to create new techniques from scientific discoveries that we thought it was regulation; in reality, it was a novelty. Even today, after four centuries of scientific development, the majority of inventions are the result of an interweaving of scientific knowledge and know-how in response to a specific problem. Understanding the difference between the process and what it ends up producing, and then the fact that what is obtained can be displaced, leads to the definition of what in evolutionary biology is called an "exaptation". It is the different use of an organ in relation to the cause of its initial selection (Gould 2002). Feathers are selected for ventilation; they end up allowing birds to fly. The jaw bones of reptiles evolve into the joints of the inner ear, etc. (Gould 2002). Examples could be multiplied.

7.3. The use of a tool: a selective system

Under this approach, the use of a tool can also be used as a selection system on the hand that uses it best, not the hand that adapts to the use of the tool. The thumb will become longer and stronger over a succession of generations as a result of selection from a population, as a consequence of the importance of using this function for survival; this variation will be selected over time and over successive generations, especially since there is already a high variability in this regard and the new tool promotes survival. For our subject , this means that it is not neoteny that leads to the creation of a culture and a compensating technique; it is the technique and the culture that creates the conditions for selection in the duration of a neotenization (which is then creative).

However, according to Lazarus Geiger (quoted by Ernst Kapp (2007)), a tool activity is always described by a term whose primitive root designates a human organ; for example in Germanic language, *mahlen* (grind), *Mühle* (mill), *Mola* in Latin, μύλη in Greek, come from the mal-mar root, frequent in the Indo-European linguistic family, which means "to crush with the fingers" or "grind with the teeth". That the instrumental activity should be named after a simple, physiological activity is universal. According to Geiger, the term that describes physiological activity existed before man used technique: language would therefore have appeared before instrumental activity.

Indeed, it is clear that cereal milling was introduced long after the appearance of language, and this is true of most of the recent tools still in use. Cereal milling would have been established a few millennia before the systematic milling activities in the Neolithic period, that is to say at a time when humans spoke languages as complex as ours. The bow itself was probably not invented long before (20,000 BP), while the propellant was invented long before and was supplanted by the bow almost everywhere except Australia. However, both were invented after language. Most current techniques and tools do not refer to tools of the Middle or Upper Paleolithic, but rather to tools of the Neolithic or even later. Who practices the ancient stone-cutting techniques?

What about an activity like cooking and all the words related to it? The root would be PEKw which means "cook, mature". Here, the root term refers to both the technique and a natural phenomenon. However, fire and cooking were invented long before the emergence of the complex language of the *Homo sapiens*; this is hardly up for debate. This could mean that, in order to teach and transmit a technique, simple terms close to the daily reality of those who are learning must be used, and that the technique was only the prerogative of a minority. Reference to lived experience may be present even in technical words describing practices prior to language. We can see that in naming a tool, we go through what is immediately sensitive. we only have to analyze the words of modern physics to see that, in fact, this was the case at the time of its development: for example the terms "force", "work", "weight", "mass", "electric current", "resistance", and recently the "colors" of quarks. If language, in its modern description, appeared approximately between 500,000 BP and 70,000 BP, it means that at least 60,000 years elapsed between the emergence of a "modern" language and the beginning of the Neolithic period. The pace of innovation over 50,000 years has been such that we can indeed imagine that most of the tool techniques described by the language were invented after language, while the others have gradually disappeared.

Processes such as cooking and baking have remained stable and language reflects this. There may be another reason. It is conceivable that at the time of the emergence of language, humans perceived the bodily basis of the technique. Any technique, even one that had been mastered for a long time, had to find its bodily referent; we have just seen the equivalent in recent physics. We understand Ernst Kapp's point of view better. Everything happens as if the technique is perceived as being based on the body's experience. Even scientific concepts are based on the same observation.

7.4. From tool-based technique to body-based technique

This is what emerges from the reading of Leroi-Gourhan's work; the technique can be conceived as an externalization of bodily functions. The first pebble to crack a nut replaces the teeth. This goes beyond outsourcing, because the action is more efficient and the risk of breaking a tooth is less. But we can't ignore the fact that this manual, non-jaw-breaking activity changes the body's requirements. In the long run, the externalization of functions by the tools has the consequence that the body itself is transformed, by genealogical selection, so that it is optimized in the use of these tools. It is an externalization which leads, in the long run, to a complementary internalization which improves the efficiency of its use. In this particular case, the selection pressure was built long before the appearance of the species *Homo*.

The fact is that we internalize the tools that we use; conversely, we project our perception on these tools. Any car driver feels the engine and perceives a skid: he is one with his instrument. Different experiences show that we feel the end of a prosthesis. Airplane pilots regret digitization and autopilots, because they "no longer feel" the contact of the wheels on landing. Our perception goes far beyond our bodies and we perceive through our tools (Merleau-Ponty 2018). To master the use of a tool is to perceive through it. This is what Simondon exposed in the use of a hammer, which is also a sensory organ.

Modern language often uses technical metaphors or analogies to describe mental processes or states. Simondon notes that Descartes' concept of logical sequencing is based on the chain concept. Today, we speak of "brakes to innovation", "getting off to a flying start", "cutting some slack", "hitting the accelerator", etc. For two to three centuries, or even more, instead of the body serving as a metaphor for the technique, it is the technique that serves as a metaphor to describe the body or the psyche. The whole body is compared to a machine and, today, the brain to a computer. Is the link to the body transformed with modern techniques?

Let's follow this sequence: instrument, tool, known as "ὄργανον" (organon) in Greek and "organum" in Latin; we will end up describing in biology a set of tissues, generally well differentiated, contributing to the realization of a physiological function; from experience, we note a multifunctionality, like with a tool; at the same time, or almost, Leibniz conceives the organism, a unified living whole, composed of organs; Kant, in *Critique of Judgment,* calls living beings "organized beings"; the concepts of organization and organism will be used in politics (see the American constitution), economics and sociology; the analogy continues and returns to its origin as in French there is a "press organ" (media outlet), a "transmission organ" (transmission device) for a vehicle, etc.

Our argument here is that the bodily origin of the technique has been referenced in relation to the body for a long time. From an identifiable time, its reference to the body diminishes, to the point not of disappearing, but of being reversed with the appearance of machines. We cannot have the same relation to technique when it is conceived according to bodily functions, or when, on the contrary, it is the body that is conceived according to technical functions. Reciprocally, isn't it fitting to conceive the body according to technical functions, if technical thought had to conceive of itself at the beginning according to its relation to the body? Here we touch on an unthought element of the technique that we will try to grasp further on. This inversion of the body-technique relationship, a major exaptation, seems fundamental to understanding the transformation of modern times, even if it was seen before the development of industry, and probably agriculture.

7.5. New evolutionary narratives

By putting aside teleological reasoning and Lamarckian concepts of cumulative adaptations, we can somewhat change the supposed order of the succession of events and confront them with archeological, paleontological and geological archives. In this Darwinian framework, we will have to think differently about the subject of selection: who or what is selected and who or what participates in the selection, and how? This question, which Stephen Jay Gould (2002) has shown to be fundamental, is central to the issue of "technique and hominization". By using Lamarckian reasoning, which is often a way of placing teleology in evolutionary terms, we can miss the key question to ask paleoanthropologists. Since Lamarckism is our most intuitive, but mistaken, approach to evolution, this may lead us to ask the wrong question rather than question from the facts. Let's give some examples.

When Darwin wrote that "we were willing to admit everything except that we started out on foot"[2], he said that we were ready to accept a Lamarckian evolution, or, better yet, an intelligent design (starting with the head), or – why not? – a creation *ex nihilo*. When we already have an idea of how we should have evolved, we have a teleological understanding of evolution; we is, at best, Lamarckian, at worst, creationists. Yet how would a thinking head lead us to learn to run? How can we undertake something based on a concept? Isn't it usually the opposite? A standing posture, especially a wobbly one, suggests an evolutionary process. Darwin understood that verticality should be the first step, because he admitted that the conceptual stage came last, which is what appears "reasonable"; favorable selection conditions were needed, which could be brought about by verticalization.

2 Quotation taken up by Pascal Picq and Yves Coppens (2001).

Simplified presentations of human evolution always remain teleological – evolution of vertical stature, evolution of the brain, the head, etc. It is a directed and continuous evolution in one direction: in other words, orthogenesis, on which the evolution of the horse was modelled. It has since been discovered, with regard to the latter, that it was more complicated; evolution had several branches; the last lineage that produced the horse was pre-selected on the American continent, before ending in the northeast of the Eurasian continent. It is not certain that it would have continued for a long time, whether humans had not selected and domesticated it.

As Charles Darwin wrote, it is the lack of archeological documentation that leads to the possibility of conceiving a so-called orthogenetic evolution. When we have only two points, we can draw a straight line directly. It took more than a century to conclude that *Homo neanderthalensis* and *Homo sapiens* are two different species; they are the result of about 500,000 years of independent evolution, albeit capable of interbreeding, and have coexisted for a few thousand years on the western part of the Eurasian continent.

We are discovering that at the time of the emergence of the modern species, *Homo sapiens*, several other relatively close species – *Homo erectus, Homo neanderthalensis, Homo floresiensis*, *Homo denisova,* etc. – were still living, and that crossbreeding may have taken place. Since Darwin, scientists have gradually understood that a differentiation generally has several branches when the corresponding biotope is scattered and segmented over large areas. Teilhard de Chardin (2008) had already noticed this phenomenon of a family tree with multiple branches for homininae.

We find many species of Australopithecus, then many species of pre-human *Homo* (considering that a "modern" language is the mark of humanity), and several species that can be qualified as human. It is because the ecological needs are very similar in all these species and because intra-group cooperation dominates that only one species has remained, capable of spreading in all planetary biotopes. There has been no evolution directed towards humans for seven million years, only an original selection pressure that led to the same biotope: everything that is habitable by humans.

The relationship between intelligence and tool-handling is another example of a general tendency to reverse terms. Knowing how to design and use a tool can be conceived as proof of intelligence; the inference is that it takes intelligence to manipulate tools and that humans make and use tools because they are intelligent. (Neo-)Darwinian evolutionary reasoning reverses the proposition. It was the first tools that allowed the selection of those who were most capable of using them at best, or even of designing new ones, that is the variability of intelligence – an empirical fact – allowed a selection based on the learning/use of already invented and available tools. We are thus dealing with a human-tool co-evolution, with a particular effect of this co-evolution: a human evolution oriented by and for use, then manufacturing, as well as tool design, by positive feedback. There is no synchronicity, but strong interactions over time.

Each biological evolution that allows for better use of a tool or makes it possible to manufacture a more efficient tool will be selected, because the group that has adopted it will be more adaptable to various environments. Here we can take up the Simondonian line of thinking and design the first tool or elementary artefact as the starting point in a metastable system (with the image of the starting point of a crystal in a saturated solution). Indeed, the use of the tool, however primitive it may be, leads to access to more energy in the still saturated pre-individual environment. The tool that allows access to the contents of any nut amplifies the resources of the medium for the one who knows how to use the tool at a lower cost. If this use is transmitted and learned, this learning ability can be selected.

According to Jean Piaget (1978), behavior is one of the drivers of evolution. It produces an outcome that may be important in terms of survival; in this case, any variability in the behavior, or internal structures favoring that behavior, will allow the selection of variants favorable to the benefits of the behavior that is both selected and a source of selection. Living in the open sea and seeking to evolve at different depths favors the emergence of density control processes to descend in a stabilized manner to different depths. There have been several responses according to lineages; sharks, teleost fish and cetaceans have found three different solutions; at the same time the range of depths by species has increased. Similarly, between flying fossil reptiles, birds, mammals and insects, adaptation to flight took at least four different routes. Around the separation of chimpanzees/bonobos and homininae, that is when bipedalism (Australopithecus or other related species) appeared, it is possible that it was a specific verticality that was imposed associated with behavior and learning, as we will see later.

The Lamarckian bias, which can be called the teleological bias, stems from an understandable prejudice. When people began to think about themselves, they were astonished by this capacity; they believed that all living people, and even inanimate objects, were endowed with it. They took this ability as their starting point. Everyone conceives the world according to the structural and ontogenetic data that has made it what it is; this process has, in a way, become an "engram" in our system of access to the world. As soon as we conceive the world, we are endowed with this ability, all of which suggests individuality. In the process of thought development, the ontogenetic and then ontological approach has thus taken precedence over phylogenetic thinking. The capacity for intentional action has been perceived in living beings according to our new mode of perception, structurally modified by our willfulness; even the understanding of physicochemical or telluric phenomena has been affected by this new capacity to access the world. It is thus understandable that animism was the first form of thinking about the world (Descola 2013).

This bias may be more subtle, since it has gone as far as the recapitulation theory: ontogenesis would summarize (or recapitulate) phylogenesis[3]. Since this Haeckel theory, it has been shown that ontogenesis can take up the paths of phylogenesis when it works. When it is cheaper, and therefore selectable, a reconstructed ontogeny leaves behind parts of the phylogenetic recapitulation. Over time, less and less recapitulative ontogeneses can be selected if they are less costly or have all sorts of advantages. Little by little only disparate traces remain, which are not a summary but rather a clue for researchers. This approach to ontogenesis means that it works like phylogenesis, it has Darwinian aspects (Kupiec 2012). The recapitulation theory can be conceived as an intuition that there is something similar to phylogenesis in ontogenesis. This can be accentuated in philosophical ontogenesis, focused on the logic of being. According to the same Darwinian process, ontogenesis acquires a form of autonomy and "substance", partially, the inscription of the history of phylogenesis. The phylogenetic traces are found in ontogenesis, increasingly attenuated, because ontogenesis has its own logic of variation-selection that allows certain historical aspects to be "shunted", gradually erasing certain traces of phylogenesis. We all live it: the past constrains us, but we sometimes try to short-circuit these constraints.

3 Formula of Ernst Haeckel in his book *Generelle Morphologie der Organismen* in 1866, found on the Internet in English. He also called this theory the "fundamental biogenetic law". His basic book has not been translated into either English or French. Several popular science books, which are very "artistic", are available.

This theory of recapitulation, like that of the neoteny of the human species, is an intuition that has made it possible to understand the processes. Nevertheless, in both cases, it is the Lamarckian prejudice that led to the need for generalization according to an intangible "law". We must take the recapitulation or neoteny according to the phenomenological approach, remain empirical, and not believe that we have found the law that would explain all human evolution. Human evolution, as we shall see, seems to be built on a logic of interaction that opens up what is possible and allows us to make conclusions only in the aftermath.

8

Animal Behavior: Hermit Crabs and Their Shells

8.1. The hermit crab: a strange crustacean

It is a shame that Stephen Jay Gould, the outstanding scientific storyteller, did not take an interest in hermit crabs – animals that are so original that they tell us something about who we are. Gould would have seen the relationship between humans and neoteny. The hermit crab's brain is that of a crustacean, with a large olfactory lobe and non-reflective detectable abilities; hermit crabs systematically use a tool, an empty shell, as a "house". They don't make it, but they are capable of change, sometimes with collective strategies that necessitate real social interactions requiring, from at least one of them, a form of generosity, which humorous scientists have called "vacation chains". A good number of them, of different sizes, find themselves around an empty shell adapted to the growth of the biggest; the latter leaves his own and settles in this new one, leaving his shell available to the next one, and so on. In the end, only a very small shell remains – probably easier to find than a very large one (Rotjan *et al.* 2010).

They can live, it seems, at least until the age of 30. Most hermit crabs are aquatic (salt or freshwater); some can live in a moist terrestrial environment and even climb bushes or trees to graze.

Hermit crabs are part of the phylum of arthropods (such as insects or spiders), the sub-branch of crustaceans, the order of decapods such as lobsters, crabs and crayfish, and belong to the superfamily Paguroidea. I don't know where the taxonomic organization of the Paguroidea is today, because there is a kind of surrealism in the one that comes to us from the 19th Century. The first family includes the Coenobitidae, with the coconut crabs and the "true" land hermit crab, whose genus is *Coenobita*. The genus *Coenobita* is taken from the name of these Christian monks at the origin of the community orders, as if the scientists who gave them this name considered them to be social animals, which led to Apollinaire's famous mischievousness on "Les Cénobites tranquilles" (*Calligrammes, Obus couleur de lune – Du coton dans les oreilles*).

The second family includes those who have been called the Diogenids, like Diogenes, the so-called cynical philosopher, living in his barrel. The mischief here is that they are left-handed hermit crabs. Was Diogenes left-handed? Since then, it seems that most of the pagodas, whose name only refers to the horn shape of the abdomen, are left-handed, even if there are some that are right-handed.

8.2. The hermit crab: an oblivious technician?

The major characteristic of hermit crabs, terrestrial or not, is that their abdomen is soft and not protected by an exoskeleton typical of a crustacean. It is for this reason, it is said, that wearing a shell allows them to protect their abdomen from drought and especially from predators. In the same way, their left-handedness is more effective in defending themselves against those who would like to take them out, a little like the tennis players whose domination of left-handedness is well known. Obviously hermit crabs have not had philosophers to speculate on their first and second nature, the first one "entirely naked and poorly protected", the second one looking for a house with a size adapted to their growth, because the neoteny of their abdomen did not lead them to this capacity: had their brain been in the abdomen, the whole organization of the world would have changed. They would have tried to build their house and would have gone to conquer the world.

Nevertheless, claiming that the search for empty shells is salutary for our animal, which has become so human, is of the same order as claiming that human neoteny has forced us to make tools. The "choice" of the first little decapod to hide in a shell resembles that of the first superior primate to try to stand upright, although the causes or reasons for doing so are certainly not the same. The consequences are also different, because a social superior primate is not a crab; their very distant common ancestor lived more than five hundred million years ago. They have had time to diverge.

9

Prejudice About the Priority of Values

9.1. The human sense of morality: an exception?

In order to compare or differentiate between humans and animals, it is the moral criterion that has seemed the most important to the majority of thinkers on all sides, from the birth of philosophy to modern times (Simondon 2012). But, as Frans de Waal (2005) points out, this is because they did not know about chimpanzees and bonobos. Nor did they know gorillas and orangutans. Today, it can be said that everything that is related to empathy and associated feelings (emotion, shame, humiliation, pride, aggressiveness and reconciliation, sense of power relations, negotiation, domination and submission, altruism, respect for the little ones, cooperation, sense of justice, etc.) is present in higher apes (De Waal 2005). The beginnings of these qualities can also be found in many mammals. It goes without saying that they cannot show that they are universals beyond them: they do not speak and, it seems, do not have the necessary representational abilities. That these feelings and emotions are more differentiated in humans is obvious, but it is a developmental process that should be understandable, if not explainable, as part of a more general process of differentiation, which is primarily cognitive and will apply to the pre-existing emotional and affective sides.

This is not to deny the real qualitative difference between a chimpanzee or bonobo and a human on a moral level. The point is to show that this difference is more about the cognitive aspect of human morality, which will make it possible to better integrate these disparate qualities. The fact that the human cephalization index is almost triple that of the chimpanzee's does not seem to be compatible with adjustments that would only concern domains that are already widely developed in our faraway ancestral lands; new qualities have appeared; they are much more cognitive than affective-emotional in nature. Alain Prochiantz (2012) describes the human brain as monstrous. How did this come about? It is hard to imagine how qualities such as those found in politicians, scientists, thinkers and inventors, writers, musicians, dancers, choreographers, painters, sculptors, poets, architects, sportsmen and women, filmmakers, entrepreneurs, adventurers, inventors, craftsmen, etc., in short, all these human aptitudes that are so varied and expressed at different levels in all of them, that collectively make humanity appear so different from any other species, even a close one. Understanding the process that has allowed such emergences to occur would be a definite step forward in our understanding of what it means to be human. This cannot concern the affective or emotional aspect of ethical or moral life.

9.2. Prioritizing cognitive ability in human characteristics

It appears, and we will come back to this, that the central difference, between humans and the closest great apes, for all that is mental, or psychic, would be the cognitive aspect: learning capacity with unknown limits, aptitude for precise copying and mime, reasoning, inner representation, inner selection of thoughts, intelligence of processes, anticipation by projection of dynamic processes (thinking according to the *a priori* of space and time), understanding of links and relationships, capacity to put a complex succession of events in narrative, sense of spirituality. Philosophers have overestimated the ethical aspects, precisely because humans can represent and talk about them, which our cousins are incapable of doing. In their defense, it must be admitted that knowledge of chimpanzees and bonobos is very recent.

One can quote Bergson with his approach to *Homo faber*, and especially Simone Weil (2001, pp. 88–105). She notes that with matter, no negotiation is useful: it is useless. It is in the understanding of networks of necessities that the characteristics of the human being lie more than in the negotiation, cunning or strategy already found in the chimpanzee, the wolf or the shark. This characteristic is still insufficient judging by the remarks of current thinkers on the crisis facing humanity in the 21st Century. André Lebeau (2006, p. 257) states: "Nothing, in the most distant past of the species, has prepared it for the problem it faces, the encounter with an antagonistic force that cannot be overcome and cannot be negotiated with. Only its recent and disparate cultural background, however unique in the living kingdom, can inhibit the atavistic behavior of dominance that leads to its loss."

It can be deduced that what he describes as the atavistic behavior of the species is of the order of the superior primate, and that in what we face we should then express our full, cognitive humanity in its least restricted sense. Is this cognitive side fully developed? Hessel and Morin (2012) suggest that it is not: "The fact that this complex character of the planetary crisis is generally ignored indicates that the multi-crisis is also cognitive." Paul Seabright (2010) shows that the transition to the "society of strangers" took place during the Neolithic revolution and that the necessary human aptitudes were already in place, aptitudes that implied a particular cognitive posture and, among other things, a wager on trust made possible by the construction of institutions that gradually allowed this trust to be established between strangers, facilitating cooperation that was generally mutually beneficial. We see our cognitive abilities only when they are in a situation where they can be used. Similarly, the technical abilities of the bonobo only appear when it is immersed in a human environment.

We differ from the species closest to us by the cognitive functions that are entangled with functions originally affective, moral, or even political (shared with bonobos and chimpanzees), and yet, as a consequence of our cumulative multi-millennia action, we are facing a new situation requiring a higher cognitive level. Maurice Pradines (1946) analyzed that the construction and use of technical objects requires symbolic thinking. Technical intelligence is characterized by "the effort to understand the reason for things that are the object of attention" while "the animal understands the meaning of things [...]. A given phenomenon can be a sign, an opinion [...]. He does not seem to try to understand why things happen".

For Pradines, "the technique is the art of doing things according to their reasons as guarantors" (Pradines 1946, pp. 21–22). It is necessary to be mobilized with a goal in mind, which the chimpanzee or bonobo and other more or less distant species already know how to do, and at the same time understanding the means, meaning and reasons for it, and to be able to transmit them. It has been shown in the "language" that a chimpanzee or a bonobo is able to learn; it is able to ask with words, but not to say the reasons. For these requests, he already has the means to make himself understood without using codified language. The relationship between the word and the thing does not escape him; on the other hand, learning a word that expresses a reason or a necessity is impossible for him. He therefore has no need to speak at all, since he would have nothing more specific to say than what he can communicate without language. He can use signs constructed by humans and practice proto-language, but he can also do without it. Talking is not just communicating. As Jean-François Dortier says, even when it has learned the signs of language, a chimpanzee has nothing to say, because it cannot talk about links, relationships, conditions, reasons or purpose. This is what Michael Tomasello (2010) describes, more than 50 years after Pradines: monkeys, including the most advanced ones, do not "see underlying causes and intentional mental states". Pradines had understood this without knowing that, precisely, this essential capacity was not acquired by the monkeys themselves, even the superior ones. And only humans make and use tools over and over again. Since when?

9.3. Role of technique in the emergence of language

Once again, the point can be reversed: it is in a context of increasingly complex tool use and manufacture that verbalized symbolic thinking, that is language, is favored. When there is technicality, there are precise conditions for the selection of a possible language which will in return allow for increased technical diversification. It can be hypothesized that if there is an identifiable break in the temporal succession leading from the first hominids to the present man in and through sociotechnical diversification and its easy adaptation to different environments, this break can be identified with the emergence of language, provided that there is no neurobiological blockage. We can see that this leads to a late model of the appearance of language in its present completeness, and in any case after the constitution of all the foundations, including technical ones, making language possible.

Language would thus have appeared as a new skill in an already technical and socialized world, amplifying the possibilities of adaptation to new situations. Humans of the Mousterian period, from 100,000 BP to 30,000 BP, are *sapiens* and *neanderthalensis*. Most of the lithic production of the Mousterian period is even attributed to the latter. The so-called Levallois method is the most advanced technique. The craftsman must preview the desired shape and master the technical gestures of striking. These are the conditions by which language can be selected, as it amplifies the transmission of know-how. This is even more accentuated for the Châtelperronian, from 35,000 BP to 30,000 BP, characterized by blades with a curved back, a culture attributed to the first *sapiens* of Europe (Cro-Magnon) and to the last *neanderthalensis* (site of Saint-Césaire). With this in mind, we can conclude that the Neanderthal man must have spoken in a way that explains reasons and motives; he was capable of mastering the Levallois technique.

This means that the demanding conditions at the physiological and neurological levels, leading to the possibility of language, must have arisen long before and for other reasons. Imagining a co-selection to speak, the control of the opening/closing of the glottis and movements of the hyoid bone to emit modulated sounds, the control of breath, controlled movements of the jaw, lips, tongue, throat, the semantic processing of these sounds and a representation of what is said, is to reverse the order of appearance of the mutations. These changes took place, in fact, in specific sequences, more or less independent, according to selection criteria that have nothing to do with language. The use of the body's means of speech was able to converge and the language of the hands, eyes, and more generally of the body, with the production of approximate sounds, was able to switch, progressively and late, to that of articulated sounds, because it had become physiologically and neurologically possible.

If the cognitive aspect appears to us first, ontologically speaking, it is because in an evolutionary process, everything that is informational (perception, information processing, imagination, representation, etc.), even if the corresponding organs develop last, takes control of the organism as a whole, since they are functions that continue the individuation process. The invention of the head, to use the expression of a one-day conference at the ENS[1] on December 7, 2002, is the invention of a central organ towards which, during the evolutionary process, all information processing converges and by which all orders of action end up passing. A decerebrate frog can still live for some time, as can a bird for a smaller time, but not a mammal, let alone a higher primate. This means that the cognitive and the universalization of thought, the speculative foundations of morality, have been built on the affective-emotional aspect already present for a long time. It has diversified thanks to the intertwining of the cognitive, which was developed later. Universalist morality is an emergence consecutive to that of

1 *Ecole Normale Supérieure*, one of the highest schools of higher education in the French education system.

the cognitive, the verbal, on an emotional and affective basis close to that of a chimpanzee (Kaplan 2001).

There is another reason to intuitively grasp that the cognitive requires the affective-emotional and not the other way around. The affective and the emotional elements facilitate survival; it is enough that there are automatic processes of survival (eating, fleeing, attacking, reproducing) that they mobilize. In contrast, the cognitive, without the support of the affective, is suicidal. It does not provide an interest in living. What is interesting or attractive is not definable from the cognitive; interest is never intellectual, it is affective. Interest is what brings, or is brought, which requires some form of adherence and inner resonance. This leads to the interesting which is a source of motivation. Hence the concept of interest rate, a financial concept rooted in affective problems. Interest, when the cognitive grasps it, takes on proportions such that it would be appropriate to look into this psychological interaction.

10

The First Phase of the Hominization Process

10.1. The conditions of access to humans through technique

Can we imagine the first technique which led to the Homininae? The general hypothesis we propose here is both sociotechnical and organizational. It is now a question of reversing the reasoning, of not considering the purpose, estimated or recognized, of a behavior or a capacity as the efficient cause of what makes them possible. It is necessary to respond both to the requirement of a necessity and to the improbability of a bifurcation event, because every phase that seems crucial is marked by the seal of contingency, if not improbability. This is not unique to humans; it relates to all living things.

Chimpanzees have technical practices, learning abilities, and the transmission skills, to pass on what they have learned. These capacities have also been found in many other ape species such as baboons and macaques (e.g. Kawai 1965; Cambefort 1981; Yamagiwa 2010). There are enough publications of systematic observations that this can be taken as a given. The first groups of Homininae to diverge from the Panina already had a technical practice, albeit limited. It is worth reflecting on the difference that accentuated the technicality of the Homininae and not on the original technicality, which can be compared to that of chimpanzees and other monkeys as technically advanced. What may have led to a kind of technical improvement and continuity, previously discontinuous and weak on the scale of one generation, and gradually amplified in the duration?

We will start from the hypotheses of François Sigaut (2012) to construct the model presented here; he laments the situation which continues to exist, as a sort of philosophical half-desert, regarding the role of the technique in the history of human evolution since the first Homininae, and opens up many paths for reflection. He seems to ignore neoteny or neotenization, but for those who are familiar with this problem, it appears as a watermark. He recognizes with Bergson that human intelligence is first and foremost an intelligence of doing. *Homo* (man) is *sapiens* (wise) because he is first and foremost a *faber* (creator). Simondon, whose Bergsonian filiation is no longer to be proven, is consistent with this idea when he says that the first philosophers, the Ionian pre-Socratics, were technicians. But not every technician is necessarily reflective; however, without a model of technical origin, how can we begin to think? The question becomes: what level of technical complexity is so problematic that one has to think about the reasons for the technical models one uses to transmit and socially preserve an invention? At what point does technique require relationality, cooperation, pre-language, protolanguage and finally language?

François Sigaut notices that technique is boring. Indeed, it is arduous, learned by trial and error and requires a confrontation with the subject "without indulgence or betrayal" – to paraphrase Simone Weill. As Einstein wrote: "Nature is complicated, but she doesn't lie." In this sense, playing a musical instrument professionally or becoming a professional dancer is boring for someone who is not in tune with what music or dance conveys, or who is in tune with it but finds the work involved too demanding and excessive. Boredom of technique simply indicates a lack of interest in the technique itself. But if what it achieves is perceived as interesting, then the actor "forgets" the boredom of the technique. I hated the biochemistry labs at university. But I loved genetics. And I initially worked in molecular biology, forgetting that in practice it was biochemistry. I hated music theory and forgot how boring it was when I played a musical instrument. The interest in the reasons for the technique made me forget the technique and how boring it was.

Here we find the opposition of the two cultures of Charles Percy Snow (2018) in which the humanities find scientific concerns boring and vice versa. It is the confrontation with this boredom or difficulty that can change everything: Jacques Boucher de Perthes, founder of prehistoric science, was able to demonstrate in around 1860 the existence of prehistoric man by taking theoretical arguments out of the box. He accepted that he would confront the material, boring reality, with the support of theory which allowed him to identify what he was looking for. Technique is boring and arduous, much like the sciences and techniques necessary for the arts, for artists. Many musicians don't like music theory either, at least in the beginning. It is the power of desire that enables us to face the initial difficulty.

Why is it so exciting to study the technical gestures of animals when we find it so boring to study the gestures of humans? Because nothing, for a human being, is more banal and day-to-day than human technique. To rediscover its magic, it takes many days, and even years, of systematic and arduous involvement. One must add the interest in the means to the interest in objects and technique, and therefore the underlying reasons that lead to these objects. Technique is all the more boring because it is omnipresent in human life. We do not know how to take the specific problem of technique in what makes us human, since we use it constantly, unless we become a technical inventor, which is certainly more fun, but remains difficult and forces us to grapple with the resistance of matter. The inventor thinks he knows the reasons, he is most often able to describe what his invention allows and how the reasons for it are new; but he is often mistaken and understands it only later. Without passion in "action" there is no technique.

Wouldn't that also be the reason why the humanities are boring for researchers in physical or life sciences? It involves studying, or even deconstructing, what for them has become "self-evident", as have most basic human techniques. The researcher in the physical or life sciences or the inventor of a new technical object, on the contrary, considers their reasons to be non-deconstructible in psychosocial terms, or, at the very least, considers this deconstruction to be pointless, because their interest lies in what they have discovered or invented. It does not matter whether the reasons are false if the results are interesting (Nietzsche). Technique can again become a human science, and to this end, the philosophy of technique must focus on the human element in artefacts and what made us human. It must combine either the two sources of boredom or the two cultures. The study of technique may become less boring for some, once it is accepted that it is one of the early foundations of humanization and that the understanding of how it works becomes a criterion for understanding what guides us, without our being aware of it.

Taking our prehominid ancestors as a basis, François Sigaut posits that four psychological and behavioral aspects of the human species, interwoven into the action of tools, must have been the result of mutations selected to allow this action:

– shared attention, which must be focused on both the means and the end, of which the great apes would hardly seem capable;

– the pleasure of success, which is not only the satisfaction of obtaining a result, nor the pleasure of the reward known to mammals and birds, but the pleasure of know-how, a pleasure which is also the answer to a recognition from others;

– the shared experience, which allows it to become transmissible and which goes hand in hand with the pleasure of success magnified by sharing;

– exchange between the sexes, because in any accessible human culture there is a sharing of techniques used between men and women, and there are no sexually marked techniques. It is an afterthought that a social construct has distributed technical skills according to sex, which gender theories reveal. This distribution corresponds to a kind of optimization of the division of labor that is not inscribed in the biology of the human species – rather in a given historical socio-technical context that has proved more competitive.

As far as the sharing of attention is concerned, there is no guarantee that the great apes do not already have an embryonic capacity. Frans de Waal (2005) references a chimpanzee who is able to prepare a pile of pebbles, well in advance, with the aim of throwing them at zoo visitors. So it becomes difficult to follow Sloterdijk (2010) completely when he writes: "the pre-human, as a thrower, operator of the shot and dismantler, is therefore already a co-producer of the forest clearing", unless he recognizes a pre-humanity in the chimpanzee that Frans de Waal is indeed willing to concede to him. Learning how to throw stones for a specific purpose, which is not out of a chimpanzee's reach, although a little awkward, could be the first experience of the pleasure of success.

Chimpanzees are less able than humans to throw a stone in the precise way; playing darts is frustrating for them. However, if the proportions are right, a chimpanzee can ride a bicycle and enjoy it. Films about chimpanzees show that chimpanzees can congratulate themselves on a successful collective action. If we could find a primary technical practice that requires attention sharing, participates in the pleasure of success and is independent of gender, there is a good chance that it could be a foundational process, provided it is in the lineage that leads to humans. Otherwise, it will have to be conceded that pre-humanity already begins before the separation of Panina and Hominina. It seems that the pleasure of success would be evident in the dog. In any case it seems beyond doubt for those who have made the effort to train them, if I believe what friends who own dogs and are attached to breeding them have told me. It thus remains to define a quantitative measure of this pleasure, probably variable and therefore selectable if this pleasure leads to the use of a competitive practice.

If these specific traits described by François Sigaut are necessary for the technical act and are present in humans, yet aren't so apparent in proto-technical higher primates, it is because they have been selected, or at least not counter-selected and associated with selected traits. At this stage, we don't have to "explain" whether there are any mutations, if they are recognized as necessary, only to show how they may have been selected or not eliminated. To the question "do they produce, from the beginning, sufficient advantages to pass the selection test?", François Sigaut answered: "Nothing is less obvious." An evolutionary biologist will answer: either your identified characteristics are indeed homogeneous and selectable entities, in which case you must find out what selected them, as they were; or you cannot conceive of how they were selected, in which case, you must either look for more precise characteristics that can be selected, or you must question your analysis. The problem is conceptual. If these characteristics have not produced sufficient benefits, it is because they are not a reason to use the tools. They are consequences.

Processes of collecting and transporting food to be prepared, requiring bipedalism, or protection against predators or competitors by stone throwing, or stone throwing to reach small edible animals, or vertical posture as protection from a sun that has become too hot would be sufficiently favorable, under sustainable conditions, for at least the first two conditions set by Sigaut to be potentially realized.

Peter Sloterdijk identifies four "mechanisms" that would be needed:

– that of insulation, which corresponds to an insulation of the group, by creating a social space, separating it from the environment;

– that of "suppression of the body" which leads to the creation of a "world" by distancing oneself from the environment which becomes less constraining, in accordance with the theses of von Uexküll (2010);

– that of neoteny, which here means maintaining the capacity for learning, discovery and invention in adulthood;

– that of transposition, that is the ability to use know-how for different purposes.

"None of these mechanisms could, by itself, cause hominization" (Sloterdijk 2010). How, from these four identified processes, can a phylogenesis be established? How do they work in relation to each other? Do they appear separately, which are the possible condition for the others to the point of identifying a succession? Already, it appears here that chimpanzees respond to certain aspects. They have a social space, their prototechnique leads them to the beginnings of the "world", they know, in some cases, how to transpose an experiment. As for neoteny, how is it possible?

We will keep these groups of hypotheses of François Sigaut and Peter Sloterdijk here, in order to confront them with a general hypothesis that is meant to be phylogenetic and not ontological, nor even anthropological, at least in the course of basic reasoning.

10.2. Verticality as the first pre-human technical experience

In the first hypothesis we will pose here that what founds the branch-like lineage of Homininae is a particular technique that precisely differentiates the Homininae from the Panina lineage. It is so significantly technical, that even after the seven million years of evolution that separate us from the chimpanzees, we still have to learn it and maintain it. It responds to what François Sigaut calls "the pleasure of success" and the "sharing of attention", this foundation of the psyche which is oriented by the technique and allows it. In order for a real technique to be born, technical success must bring real pleasure to the point where it becomes a driving force for technical entrepreneurship. It is also necessary to be attentive to both the way of doing things and the objective to be achieved. You only have to observe a young chimpanzee trying to crack a nut to see that while the pleasure of success is a motivation that is present in him, he does everything in disorder and random repetition, without paying specific attention to the method nor its effectiveness in achieving the pursued goal. It takes years for a chimpanzee to learn how to crack nuts efficiently. The fact that the young chimpanzee doesn't do it quickly means either that there have never been any mutations that improve a young chimpanzee's abilities – which is unlikely – or that there have been no selecting conditions, that is no advantage for the future. We now know that a small bonobo or chimpanzee reveals, in a human environment, learning capacities that would be unsuspected under "natural" conditions (and vice versa). Different environments equal different selections.

My hypothesis starts from the idea that vertical posture has nothing to do with a purely biological posture in humans. It is socially constructed. It is a technical learning process, usually unconscious. This can, quite rightly, make biologists jump up and down, and they will have many arguments to assert that it does not "hold up". The agreement is remarkable between this posture and the specialization of the legs and arms, the morphology of the feet, or the curvatures of the spine, concentration of the digestive system, evolution of the pelvis, position of the head with evolution of the occipital bone and displacement of the foramen magnum, etc. This is undeniable. From Australopithecus to *Homo sapiens,* the biological evolution is remarkable, and took place over four million years.

The theory defended here is that these are precisely selected adaptations and that this is verifiable and verified. They are all the result of evolutions, at different times, during the millions of years *after the standing position* of the Australopithecus, which, it must be admitted, was not very skillful. The gait was swayed, because the morphology was only partially adapted; the Australopithecus was detected by paleontology while the vertical posture seemed to be acquired, but far from perfect. Certain musculoskeletal elements prove this, even if this morphological adaptation was far from complete. It had already taken about two million years to stabilize it without being complete. One to three million years before, *Ardipithecus Ramidus*, *Ardipithecus Kadabba*, *Orrorin Tugenensis* and *Sahelanthropus tchadensis* suggest bipedalism without absolute proof. Between the attempted bipedalism of the beginnings and the evident bipedalism of *Homo erectus*, at least three million years have passed.

We have many arguments to defend the idea that vertical posture and bipedal walking are constructs that go beyond biology alone, even today, after at least four million years of evolution during which biology has adapted and the vertical gait has imposed itself within the framework of modern human morphology and biology (length and relative forces of the legs and arms, position of the head, curvature of the spine, hip and shoulder joints, etc.).

Firstly, this verticality continues to pose problems, for example, the fragility of the lumbar vertebrae. Then, it appears that the child learns to walk, and this gradually, in the context of real efforts, most often in response to parental stimulation, and in any case with the aim of imitating. He is supported in his efforts without ever being punished for not succeeding, which is today considered the best form of teaching. Not all children learn to walk at the same speed, nor systematically in the same way. The pleasure of success is visible in the child who takes his first steps. Learning to walk is a rewarding effort, and the division of attention between how to do it and what to do seems difficult to deny. This learning is touching in terms of the effort it requires, which is vital, despite the hesitant nature of the first step; it is rewarded by success.

This does not mean that walking is optimized after a few million years. Even today, you still have to learn to walk in a different way in order, for example, to become a supermodel, or to practice hiking or mountain walking, or even collective walking in rhythm, of which there are many variations. Walking is a technical activity that is mastered unevenly and differently in different cultures, which is revealed by simple international observations. Walking, running, swimming and diving are all learning activities. Mauss and Haudricourt have shown that there are many different body techniques (Mauss 1936; Haudricourt 2010)[1]. Mauss writes: "The body is man's first and most natural instrument. Or to be more precise, rather than instrument, the first and most natural technical object, and at the same time technical medium, of man is his body." Mauss and Haudricourt show that man made his own body both the first technical object and the first technical means. The body is both man's immediate technical object and the means by which he gains access to the technicality that will enable him to develop tools and instruments. Our hypothesis is that the first step in the body's technicality that led to the human being was to learn to stand upright and walk vertically; this created the conditions for selecting a biological optimization of walking, while continuously freeing the hands.

1 The obstacles encountered by the two authors for their publications show the denial they faced.

It is therefore not, according to this reverse reasoning, only because of the tools he built that man has ended up using his body as a technical object. *It is because his ancestors learned to use their bodies technically in a context of "problem solving", also involving tools that the "technical posture" gradually developed.* From there, their descendants were able, little by little, to learn how to deploy it on objects as an extension of their bodies, especially since the first signs of the use of tools were already present: they are found in chimpanzees. When a chimpanzee throws stones, he must find a posture that frees at least one arm, and this posture may not be vertical. However, if he needs both hands, he will have to stand on his legs only. Chimpanzees, or at least some chimpanzees, can hunt with tools (Pruetz and Bertolani 2007). This seems to concern "immature" animals more so than adults.

We now know that we share many characteristics with bonobos or chimpanzees, some of which are more developed in humans, although perfectly identifiable in them (de Waal 2005, 2009, 2014). Moreover, we master a technique that we have to learn and we learn it at a very young age, just by observing children learning to walk: carrying and throwing objects while walking standing up is one of their first technical pleasures. Chimpanzees do this with difficulty, and in a risky way.

Georges Vigarello (2001) shows that the verticality of the body in humans is a representation. The straightening of the body comes from a will, requires exercises and a long apprenticeship. This straightening requires work that mobilizes this sixth sense of kinesthesia and whose organ is the vestibular system of the inner ear (Berthoz 2002), which only developed in an accentuated way from *Homo erectus* (Skoyles 2006), allowing an increase in size, then the endurance race. This verticality is being worked on and gymnastic exercises that require backward curves are impossible for other higher primates (Skoyles 2006).

If learning to walk is an apprenticeship for a small child, even a baby, made possible by the double characteristic that is "the pleasure of success" and "the sharing of attention", what can we say about such a success in our distant ancestors, before the Australopithecus? What might have led to this effort and how did it come about? Here I propose a hypothesis that may be difficult to falsify (in the sense of Popper), without being impossible; I will first deploy it to critique it and at the same time expand on its possible variations.

Small bonobos and chimpanzees are able to walk vertically, approximately until they are weaned. They lose this ability during their growth, which is morphologically incompatible with a stable vertical stature. That the infants walk vertically is not a collective construct but a simple biological fact. We can therefore assume that this was the case for our common ancestors. Those who later became Ardipithecus, Australopithecus, or other neighboring species, would only have learned to maintain upright walking beyond childhood according to a form of transmitted learning. The postural evolution of chimpanzees, under "wild" conditions, has been studied by Sarringhaus *et al.* (2014). He shows that, in their movements, infants stand upright much more often than adults, whether hanging by the hands, climbing trees, or walking. They spend almost twice as much time moving around as adults. Four-legged walking develops gradually with age. Research on the transition to bipedalism shows that specific conditions (food piles, eating high up, carrying objects, the need for alertness) are plausible processes that may have led to bipedalism in Homininae since they cause chimpanzees to be bipedal (Videan and McGrew 2002). Furthermore, the study of facultative bipedalism in non-human primates shows that bipedalism is not erratic; it is already structured (Druelle and Berillon 2014). Studies on bipedalism are numerous and show that the transition to bipedalism in higher primates can be spontaneous but also induced. Human-type bipedalism becomes energetically advantageous (Sockol *et al.* 2007), so selection pressure is possible if the behavior requires it.

Vertical stabilization of the body requires increased brain development, which is uniquely human, and appears late with *Homo habilis* and then *Homo erectus* (Bramble and Lieberman 2004; Schmid 2004). The characteristic of current human vertical bipedalism is based on a sense of balance (Berthoz 2002; Skoyles 2006). It took more than two million years to change from a walk that still resembles that of a small chimpanzee, even though the morphology has already adapted to the specificities of human verticality. This is the time it took for the brain to double in volume and for the technicality to reach a high level of specialization. Note that Australopithecus were relatively small even if the variability of their size remains high: the increase in the size of Homininae is a consequence of the acquisition of a sense of balance.

The contradictory argument that the little human child learns to walk late, rarely before 16 months, does not take into account phylogenesis; the human child appears almost four million years after learning to walk verticality; in the meantime, many events take place; he has, during the first month, a kind of "walking reflex", a trace of phylogenesis, which is efficient but fading away; walking on all fours (on his knees) before *learning to walk vertically*. As for the Australopithecus, its neoteny concerned verticality; it had a big ape brain, between 350 and 450 cm^3; it was not born "premature"; it started to walk small, like the little bonobo or chimpanzee. The neoteny of the fetus and the infant did not exist in the Australopithecus – it happened millions of years later, which postponed the learning of walking, which had already formed at least two million years before.

One can imagine a rather counter-intuitive hypothesis that sheds a potential new light on the emergence of humanity. One can imagine a group in which the little ones bring nuts, or any other product or dried fruit to be shelled or broken, to the adults to do this work. It is fun and effective. It works so well that little by little, as the group moves around, the group organizes itself with males at the periphery to protect the group (Sloterdijk's insulation), others, in the center, males or females, break the nuts and the young bring all these potential food sources they find. Gradually, a selection is made of the most active and skillful offspring. In short, a form of work or activity of the young, more or less constrained, probably playful and involving a form of competition, collectively oriented, is taking place, because they are more able than others to bring food products to break open; they can take them in their hands and move around. This requirement, which can be negative for the survival of the young, favors those who are more efficient and who know how to maintain the vertical, liberating posture of their hands for longer. This favors the "pleasure of success" and the "sharing of attention".

Let us insist on the role of the game. It is known that many children like to play so-called dangerous games such as "the choking game" or "the blushing of the face that ensues", the latter consisting of holding one's breath for the longest time possible; it is the repetition of an initiatory rite of passage found in certain African tribes. It goes without saying that if this dangerous game had been invented thousands of centuries ago and if this necessarily dangerous rite of passage had become a selection criterion, in the long run it would have selected populations more capable of controlling their breathing. As for the scarf, it would favor those who control their state of consciousness (i.e. the one that stops before losing consciousness). We can imagine such a scenario here, playful and "agonal" according to the expression of Huizinga (1951), that is constructed according to the logic of competition, with a process of selection.

In such a group, the vertical posture is imposed little by little over time, until after many generations, the children are freed from the obligation incumbent upon them, inasmuch as they all, little by little, become vertical. Because of the pleasure of this posture, they keep it and demand it from their descendants. This oriented activity of the little ones will have freed everyone, in the very long term, by producing a collective obligation to have the vertical posture which will have many consequences. There can be multiple modulations of this hypothesis. It is known that in some chimpanzee groups, immature chimpanzees attempt hunting with 70-centimeter spears (Pruetz and Bertolani 2007). In fact, many possibilities are conceivable. These relate to the environment and its resources or the reasons for the need to remain vertical for longer periods of time. As this is an indicator of childhood, other forms of competition do not take place. It is a game for children and young people that gradually enhances this state in adults. The principle: a collective demand on a learned and prolonged verticality effort, but whose success has an advantage over feeding and reproduction, whether individually or collectively. Such a model, approximately maintained, produces in a few thousand generations a population at the beginning of verticalization.

This hypothesis has four main interests: first, it conforms to a scheme of exaptation; it fits with the definition given by Nietzsche (in anticipation) and modern evolutionary biologists, that is verticality was only selected because of a form of collective behavior; its multifunctional interest will only become apparent much later. Second, although vertical posture frees the hands, it is rather the use of the hands to carry or grasp objects that has led to its preservation over time, until, after many generations under selection pressure, it is maintained in adulthood. We shall see later on the wealth of possibilities offered by this method of reversing, against our intuition, the relationship between what selects and what is selected. Third, because a practice considered *a priori* as a disaggregation form of the logics of primary biological survival could have led to a posture favoring exchanges and relational development. Fourth, this hypothesis supposes that a slight evolution of social relations, which may have depended on random events, could have led to a sharing of tasks corresponding to biological characteristics, themselves subject to selection, which in turn could have led to a profound transformation of inter-individual relations. The intertwining of the social and the biological constructs the Homininae family from a specific learning process: verticalization, which is always associated with the use of tools.

We can imagine that the groups live in an environment where small rodents or small prosimians proliferate, sometimes going to drink and in any case venturing out to where there is food to eat. The game of the small ones is to stun one of these small animals by throwing stones to them, even by throwing small spears; I myself tested this method when I was less than 10 years old and I was able to take a viper in this way, by first throwing stones at its head which stunned it long enough to take it by the tail without too many precautions and to keep it alive. You just have to know how to aim, which corresponds to, according to Sloterdijk, “already being a co-producer of clearing”.

It is certainly necessary to imagine an environment that is large enough and homogeneous enough, over a certain period of time, for such groups to gradually come into being. Nevertheless, if the feeding and predation environment is an essential condition, the hypothesis of a longer verticality of young people must be thought out according to social conditions that favor (select) a longer verticalization. If this is accompanied by the pleasure of success – that is a social aspect is involved – it is clear that, here again, it will be selected as learning, even if the population remains small. Imagine that selection takes place at the edge of the forest, with many modalities favoring the maintenance of the suspended and erect vertical posture, an orientation such as that of *Ardipithecus ramidus*, a million years before the Australopithecus (Lovejoy 2009). It is consistent that this is learning under collective pressure; it changes the conditions of existence; these conditions change the pressure of selection; the pressure of selection promotes more effective learning. It is conceivable that this selection pressure for a more sustainable vertical posture for the young could have come from even a slight conversion of the collective life, which was long enough to lead to detectable musculoskeletal changes.

Some people want to naturalize verticalization: that would be a selection of those who remain vertical because it gives them an advantage. For example, being vertical is more advantageous in a warmer, drier climate. In this context, young people who stay vertical longer die less from dehydration, and within a few thousand generations a group would become vertical. Suppose the young understand or feel that by standing upright, they are more resistant to sunlight and can take better advantage of parsimonious shade, much like meerkats that stand upright in groups to resist radiation. It is a behavior. Is it instinctive or voluntary? Spontaneous or learnt? If it is learned, learning will be the source of change, and those who can learn better will pass on that learning ability, which will increase over time because it promotes survival. There is a specific pitfall here: where is the fun in success? Meerkats that resist the sun by standing upright often fall over from exhaustion; in any case, they don't pat themselves on the back. If it's just a question of survival, without relationships, without imitation, emulation and the pleasure of success, it doesn't work as an evolutionary engine for something else, such as carrying and bringing things.

Henri Guillaumet, after a plane crash in the Andes, thanks to an extraordinary will, was able to survive and he told his friend Antoine de Saint-Exupéry: "What I did, no animal has ever done" (Saint-Exupéry 2016). Indeed, and Sławomir Rawicz, the author of *The Long Walk,* could say the same thing about his escape from Siberia, in the spring of 1942, which led him to cross the Gobi Desert in the summer, then the Tibetan plateaus in the autumn and the Himalayan range in the winter[2]. Yes, man is capable of exerting on his body a hold that no animal can imitate. He is capable of using his body, as Mauss would say, as his first technical object. But he had to learn how to do this. The hypothesis that this began even before Australopithecus is not absurd. For how could he use his body so amazingly today if it didn't come from the earliest beginnings of hominization? Our hypothesis is an application of Occam's razor.

Even a bonobo, chimpanzee, or gorilla can undertake exploits by forcing their bodies. The general public film *Bonobos*, directed by Alain Tixier, tells the story of a little bonobo orphan from Congo, called Béni by those who found him. Claudine André, who opened a reserve unique to the world, saved him and plans teach him again how to live in community. Jealous of a peer who knows how to swim easily, the young bonobo, at the risk of his life, wants to know how to swim better than his competitor, because there is a female to conquer. He is neither Henri Guillaumet, nor Slawomir Rawicz, nor Turenne or the soldier of Marathon, he lacks the mastery of effort,

2 The authenticity of the book has been controversial; it could be plagiarism and some aspects seem exaggerated. But the original story seems real. The list of such exploits, told throughout history, is very long; one can go back to the memoirs of Turenne and further to the Marathon soldier.

concentration, “double attention”, and resistance, and yet, already, he can put himself in danger for an uncertain success which he holds dear. Humans are the result of a very long selection of these abilities. Training is also a way of showing that it is possible to lead many mammals to learn skills that violate the body, even for animals that are not “voluntary” enough for this kind of torture, simply by a demanding pressure, as Skinner showed. One could apply the concept of self-domestication to this evolution that has led to modern humans.

It is not unreasonable to assume that verticality has been learned, because it still is. It is not verticality that has been preserved, over millions of years, it is the desire to be “standing up”, an expression that still keeps today, millions of years later, the immediate meaning that each of us, each individual, each ancestor of the lineage that led to the human has felt. To be standing is to be vigilant, attentive, precise; it is, for a human, “being alive”. According to this hypothesis, standing is the founding act that will then lead, without being totally determined, to the different “choices” that will lead to the human. For if there is one criterion, the oldest, which separates us from the Panina, it is verticality and this capacity for sustained attention. It is not by chance that the term Homininae was chosen to separate the true vertical lineages from the other lineages of Hominidae. How this was done is another story. A story has at least the advantage of making one feel where “the heart” of the problem lies. The hypothesis made here can be summed up as follows: the first groups in which the maintenance and pursuit of verticality during childhood and independently of gender, according to a learning process, improved survival, population growth and later the proliferation of groups, created the gradual separation of hominids.

Other hypotheses on the causes and/or benefits of such selection pressure are conceivable within the framework of a central hypothesis: the vertical posture is an acquisition resulting from an apprenticeship, which took place under collective pressure. This is conceivable because in chimpanzees and bonobos, it differs at a crucial moment: when they are small, they spontaneously stand vertically, then their morphology evolves and leads them to this very particular walk with the support of their arms on the backs of their hands. The little bonobo, at birth, has surprisingly human body proportions; it diverges afterwards. All the individuals of the three species of great apes closest to us are capable of standing for some time with or without the help of sticks. This would suggest that the behavior of our cousins so close to us could be a naturalistic regression, after the effort towards verticalization, with a return to walking on all four limbs. Our common ancestor would have stood approximately upright, the verticalization would be neo-technical, technical, and the Panina would be the result of less or different effort.

This hypothesis could be taken further and we have some arguments to support it. First, *Sahelanthropus tchadensis*, known as "Toumai man", although not a human, is a contemporary of the ancestors of chimpanzees, perhaps bipedal, but this is not certain. The quadruped walk of chimpanzees and bonobos is peculiar: they do not lay their hands flat on the ground, whereas any small human lays his hands flat. It is therefore conceivable that this evolution towards bipedalism took place before the separation between Panina and Homininae. Chimpanzees and bonobos would be branches that would have lost this learning of bipedalism, which, in the long run, selects the morphology. Yvette Deloison (2013) defends this idea in quite a convincing way; it seems possible and agrees with our thesis.

All mammals, and all the more so as they are evolved, learn a good part of their behavior; this is the case of the Panina. The Australopithecus, for its part, comes from a lineage that learned to walk vertically and has maintained its learning. It was laborious and long, of the order of two million years, and it was hardly elegant; this walk must have required a real effort which must have had advantages, or at least no major collective disadvantage for the groups having evolved in this way. This permanent tension of verticality, which we can imagine as coming from life on the edge of the forest, to be constantly watched, tells us this too. The standing being looks further away, farther away. His hands are free, he can use them better. The clearing of the being begins to open up.

We can think of this in another sense; if verticality is "natural", biological, then, how are we going to explain that the human body, at once the tool, instrument and utensil of its owner, is so technicized? How did this profound technicization of the human body begin, and how did this resistance to tension, this permanent effort, come about? If it did not begin with verticalization, when does it begin? Whatever the source of this tension, it has manifested itself from the very beginning of this long process that took more than two million years, since the divergence of the first Homininae dates back about seven million years and, two million years later, it cannot be said that their rather wobbly posture and their spinal column have become flexibly adapted to walking upright.

We can even imagine that, for two million years, groups of great apes existed with more or less marked verticality in the body, able to reproduce differently, and that it is only after a sufficient time by emulation and social mutual aid that the verticals became separated; more tonic, more adaptable and masters in stone-throwing. As for those whose verticalization has become naturalized, we have, through some examples, a kind of counter-test: they lose the ability of humans to invent their own environment. Naturalization is either the return to a form of quadrupedalism or the biologization of verticality.

There is still a philosophical question. Is this proposal not a simple displacement of the ontological question? How does the technique come about, even bodily technique? There will indeed be a need to further question this supposedly founding point, but, in fact, the purpose of this hypothesis is to answer a question. What process can be implemented, during this distant past, so that it can then be a comprehensible evolutionary engine in this evolution towards the human, which otherwise remains just an empirical description? Many researchers postulate that there is a technical-biological co-evolution. It should be made explicit: how? However, we will see later that, based on this hypothesis, this seems to be the key. It would be enough to verify that this is indeed the case. Afterwards, it will be possible to philosophize about this problematic seniority. It will be necessary to take into account that bonobos and chimpanzees also know how to play and that, depending on the conditions, they prove themselves capable of inventing new ways to play. They have strictly technical relationships with objects. They take and release; they can exchange the object. The object can serve as a mediator.

The second question, which is answered by this approach, is the question of learning selection, since it is often said that the learning that our big brain allows us to do leads to a halt in biological evolution; we will always be able to adapt without the need for biological selection. However, verticality as a permanent source of learned technical use puts in place a new relationship to the use of tools. Technical use becomes a selective process, even if, at the beginning, it may be "just a game". However, we will see in the following that the body-tool interaction allows a co-evolution because it is a learning process. We have a configuration where the improvement of the learning leads to the improvement of the tool and vice versa. Learning can be both selected and selecting. Without a permanently necessary technique, there can be no selection by and from learning; this technique, even physical, is constructed by the polarization of the biological individual and of learning that comes from elsewhere, that is learned. This is in accordance with Sloterdijk's thesis of the suppression of bodies.

10.3. The consequences of verticality

The accentuation of experience sharing is a consequence of this process of verticalization whose learning is collective, mixing competition and cooperation among age groups of the beginning of autonomy, which facilitates a shift. If this can be learned, it can be imitated and emulated. Mirror neurons exist in almost all monkeys. They are a group of brain neurons that are activated both when an individual performs an action and when he observes another individual performing the same action, or even when he imagines such an action – here in a human context (Rizzolati and Sinigaglia 2010). The effort to maintain verticality has been shared; it cannot be a strictly individual effort; it has mobilized and still mobilizes mirror neurons.

The same is true for the exchange between the sexes since this technique would have been acquired before the sexual differentiation of prepubescent individuals. If a general, non-sexual selection takes place at the end of early childhood, this makes it possible to understand the slightest sexual differentiation of humans and access to technicality independently of sex. It is a collective, made up of prepubescent males and females, which contributes to the collection capacity of food; this may require decortication, or the hunting of small rodents, or something else conceivable. The selection is identical in both sexes and they are mixed in the competition. The weak morphological and biological differentiation between girls and boys before puberty suggests that there was indeed a homogeneous selection regardless of sex in early childhood, in all aspects of verticality.

This heterochrony in development, which is selected by the social organization, concerns only the standing position and the release of the hands. It does not concern the volume of the brain. Should we see changes in the homeotic genes? The mutation supposed by François Sigaut is indeed selectable according to a conceivable process. Moreover, this selection process also favors any mutation leading to new neuronal connections accentuating the pleasure of success – we have seen that it already exists in Panina, less accentuated than in humans – and of course the sharing of attention favored by the learning of a posture allowing manual activities. It is a process of selecting, by its practice per se, that allows it to emerge. From that moment on, a new selection pressure is put in place, which will favor any cerebral revision in favor of the ability to learn.

The conditions for the selection of this posture are described here as problems of social organization, of sharing functions according to age, a form of specialization in an already social structure. Learning is shared at least within one age group, regardless of gender. These are social and organizational changes, associated with group protection and feeding techniques, made possible by a vast and homogeneous environment: moving to areas that produce food to be harvested, etc.

We have here a hypothesis that meets the four necessary criteria posed by François Sigaut for the emergence of a technicality. As soon as conditions select the extension of the duration of the vertical posture of the young, by using their natural activity of play, by taking advantage of their mobility and their capacity to respond to adult demands, it also selects the first two characteristics, in relation to the technicality of maintaining a posture. The general framework actually selects the other two criteria.

This hypothesis can be refined and also responds to part of the problem that Peter Sloterdijk had identified. Isolation, as well as the removal of bodies, is created with and through the "work" of the little ones, given that the optimization of their activity will be achieved by the general protection of the adults: on the one hand those outside the group who create the isolation, and on the other hand, inside, those who use what is brought in to make it consumable. It seems preferable here to adopt a Darwinian reasoning. It is the slight variations in the games of the little ones, combined with the protection of adults, that create, through progressive interactions, over time, an adequate structure for the children's activities, according to a playful, albeit dangerous, competitive logic. It is also possible to imagine the mobilization of children to throw stones together towards predators, a mobilization that can be experienced by them as dangerous and playful, while being of immediate interest.

This society remains that of a Panina, while being structured by a division of labor involving the little ones; we have seen that immature children can hunt with tools. Over time, this creates a selection that provokes a progressive transformation of children evolving as adults towards verticality. Make no mistake about the selection; it can be the in and/or out group; it is necessary, in one way or another, that the groups that have more young people in this activity requiring verticality are more efficient for the development of their populations. The in or out group differential must have an effect on reproduction. Here, the seductive hypothesis would be to enhance the ingroup selection; groups favoring these practices would be more efficient. However, this hypothesis implies that this collective behavior gives rise to a selective advantage. It is a selection known as "hierarchical"; it concerns groups, groups of groups, more or less important, moving, interferential populations, without forgetting that in the great apes, the young females leave the group to integrate into another one.

This progressive verticalization is a process of neotenic heterochrony. Children, through selecting their activity, reach puberty after a certain number of generations, keeping one characteristic of a child: verticality. The stone throwing becomes a consequence of this verticality, because a chimpanzee, and its distant ancestor (the closest common ancestor with the Homininae), can throw stones, but less skillfully. Verticalization, associated with the throwing of objects, as a process resulting from an apprenticeship, selecting neotenization, responds to the characteristics identified by François Sigaut and Peter Sloterdijk. It's the opening of the Heideggerian clearing or the opening of possibilities.

Of course, other hypotheses are conceivable: the exercise of confrontation with the analyses of Sigaut and Sloterdijk, keeping in mind the analyses of Pradines, the general approach to human cognition as outlined by Tomasello and the theorists of extended, embedded, integrated and enacted cognition, shows that general conditions involving selection pressure at the end of early childhood are necessary to maintain a verticality which does not depend on gender. To that end, isn't the establishment of collective emulation, task-sharing within the group, and relative isolation from the environment a necessary consequence of that sharing? This could have been put in place over several tens of thousands of generations, in several places in this great Africa, the mother of humanity. Attempts have been multiple, with the great vertical apes multiplying in diversity.

This acquisition of a verticality that is not "natural" but constructed could be lost, a naturalization could take place, and a primate that is "naturally" vertical, or returning to the quadrupedia, could take place. Our hypothesis leads to the necessity that, in the tortuous lineage that leads to Man, this naturalization did not take place. However, the available data show that there have existed several vertical "species", in close proximity to the nursery from which they came, some of which have specialized and ended up in deadlock. In fact, according to the comparison of paleontological molecular data, it seems that the separation between Panina and Homininae took place about seven million years ago, a date close to that of *Sahelanthropus tchadensis*. Over a period of time ranging from four million to two million BP, many species of Australopithecus or related genera have been discovered. Among them, it is striking that *Australopithecus afarensis* and *Australopithecus africanus* are considered to be the closest to our ancestors, *Australopithecus anamensis,* which may still be older; they are the most slender and have the most neotenic character. Implicitly, neoteny is the mark of the amplification of learning.

This means that their strength, which has enabled them to cross the millennia and give successors, comes from this fragility specific to the technical species, whose progress is not biologically assured. It is the result of collective pressure, emulation and learning, which define the technician approach. More slender, but more technical; more technical leads to more slender. That would be the secret of our origin and our evolution. This bipedalism has been written in stone. In Laetoli, Tanzania, hominid steps, including those of an adult accompanied by a child at his side, have left footprints preserved in petrified volcanic ash. In 2016, many more footprints were discovered, a veritable group of individuals of varying sizes, some larger than the skeletons discovered to date (Hatala *et al.* 2016; Masao *et al.* 2016).

That was three and a half million years ago. They weren't afraid of the volcano and they walked upright. These footprints are attributed to *Australopithecus afarensis*. But nothing definitely seemed earned; at each new step, the bifurcation was significant. Either it was regression and naturalization, or it was the continuation of the technical effort, of the construction of a protective and emulsifying social envelope, which can be seen physically by the appearance of new tools, or by the traces of these tools.

Indeed, after walking, or in parallel, according to different lineages, there was running, the development of manual skills, relational intensity, the formation of the couple, the increase in collective solidarity and the relative closure of the group, which allowed the exploration of the environment by taking in our world. Technological collective selection, from the dawn of hominization, and maintained as such, is the secret of what – in seven million years, enough to produce at least seven successions of different species – gave rise to *Homo sapiens*.

11

Towards the Verticalization of the Genus *Homo*

11.1. Aging of technical achievements

This first hypothesis of the acquisition of the fundamental functions allowing technicality to emerge from the learning of the vertical position, as well as these four functions described by François Sigaut as a condition of technicality, suggests that they were acquired much earlier than he himself assumed. Moreover, the selection conditions were produced by the collective behavior of the group, and this over a sufficiently long period of time. It remains to be understood how this acquired learning will shift in time, with a total loss of the vertical posture of the infant, through the course of evolution, since the human baby finds itself, a few million years later, not being able to walk before 18 months.

The loss of fur and sweating abilities also seem only to be biological consequences selected by a shift in living conditions. Darwin concludes that the human body could not have been "stripped" by natural selection and yet this happened long before the spread of humans across the planet. The loss of fur and sweating are consequences of the technical relationship with the body. The ape becomes naked, as Edmond Morris (1967) put it. It reveals its fragility and stirs up desire. It is difficult to tell from the skeletons when humans began to cover themselves. We imagined that it was when we left Africa – that with the help of the cold, clothing was invented; or that it was for esthetics and appearance. But we could have hardly imagined an earlier invention from 30,000 or 40,000 years ago. It was by studying head and body lice that the date could be set first at 70,000 BP and then at 107,000 BP

(Kittler *et al.* 2004), and we can perhaps go back even further, because this date corresponds to the separation of the lineages of the two species of lice, which implies more than 107,000 years without discontinuity. In short, that is at least a few tens of millennia before the visibility of the "human" revolution, in accordance with the Darwinian hypothesis. There should be a link with language, because to dress is to hide one's biological specificities, to say something about oneself, and to enter visibly into a world that affirms social relations to the detriment of bodily relations. Here again, according to our theory, it came from the little ones, moving, swaddled, protected, for longer and longer, until they were all clothed, even as adults, with the fur removed in the meantime.

This is a constant in our knowledge in the discovery of the emergence of important functions that are not easily identifiable in material terms. We position them at the measured moment of their appearance, as provided by the paleontological archives, and, depending on the various discoveries or observations, we are obliged to shift them to older times.

The first reason is obvious: it is the lack of archeological data that Darwin was talking about. A technique, or a body datum, is traced back to the earliest date of its discovery – until an older one is found. This was the case with the lungs or the feathers of the birds that we now know were developed in the time of the dinosaurs. Birds are not only the descendants of dinosaurs, they are their present-day representatives.

There is another reason. We see feathers in the fossil footprints of dinosaurs, which we did not see in the past because we did not imagine it. This was the case before the understanding of what a fossil was; we didn't see them as fossils, we imagined physical phenomena. We can't see what we don't imagine. The theory of punctuated equilibria has shown that evolution is not constant, whereas before this theory, it was assumed that progressive evolution was little seen because of the lack of geological records (Gould and Eldredge 1977). Since the theory of punctuated equilibria, multiple "stases" have been described along many lineages, alternating with periods of more rapid evolution. A phenomenon is only clearly seen when it is well understood, conceivable and identified. Cuvier already said that a good observer must be a good theorist. We see here that theory is not what is needed to set up a practice; it is what is built, step by step, in interaction with observation. A "good theory" allows for a better observation, but it can be built from observations from a previous theory.

The underestimation of how ancient the use of broken/cut stones is, both for scraping skin or bones to retrieve a richer diet than vegetable products and for breaking bones to recover marrow, has been longstanding. Recently it has been proven that this happened at least among the Australopithecus (Ferraro *et al.* 2013). We do not see why a group that mastered the learning of verticality, that is whose individuals became technical practitioners of their own body, liberating the hands and using various tools, would not continue in this direction. Throwing stones, breaking them, even randomly, using them for other purposes, rubbing, scraping, acting in a coordinated way in a group, this is what a vertical primate with a brain similar to that of a chimpanzee should be able to achieve without any problems. In doing so, it alters the selective processes of its own evolution, because those who know how to do this can transmit their know-how; and this selects, both within and between groups, the abilities to imitate.

From the first vertical primates, whether Australopithecus or other competing species, something new emerged: more active predation. It has long been claimed, and still is, that they were scavengers. Yet even chimpanzees do not hesitate to actively hunt together, at least, they are known to hunt colobus. We imagine that Australopithecus hunted in groups, even cautiously, using stones and sticks. Injuring a gazelle with stones and making it fall became a "child's game". As soon as the idea of active predation is acquired, we look at it differently. Attacking a cob in the prime of life becomes evidence of active hunting (Bunn and Gurtov 2014). We see here that the register of proof changes as soon as the theory changes, which every epistemologist now knows.

We could use Sloterdijk's expression, itself taken from Heidegger: the founding gesture, learning to remain vertical as long as possible, knowing how to carry objects or throw them, produces distance and clearing, but not immediately, because even under selection pressure, the evolution of living beings is relatively slow. Our technical account also describes that the human adventure is a technical adventure, even if it can be translated, after the fact, as that of the emergence of being, of art, of civilization. The human being who profiles himself in this act emerges from the environment to open himself to a world he invents. To begin his technical adventure, he also perceives himself as a technical object. He can then use a stone, to crack nuts or bones, to throw it towards what he intends to hunt, widening his space and creating a distance between himself and his environment.

Thus, with a brain of less than a third of that of the human brain, all or almost all is already ready to open up to the human yet to come, depending on contingent, unpredictable elements, which will be selected according to new conditions of existence. Prometheus did not bring technique to man: he taught those whose descendants would become humans to stand upright, learn it, maintain it and to use their hands and arms to make and use tools. His constant effort exerted upon himself leads him to an unceasing effort towards what surrounds him. With this in mind, Epimetheus created a primate who was ready to accept it, bent under effort, but who constantly rises again and again for action.

We can begin to reflect on the possibility that many qualities appeared earlier than imagined, are variable and were selected gradually according to what they brought to a technical application in progress. They become conditions for the selection of biological qualities identified as human, and then of qualities recognized as human and going beyond pure biology. A tension towards verticality, the corresponding effort, the pleasure of success, are already the beginnings of a way out of pure biology, in the same sense as saying that "the living comes out of matter". This "exit" is also a selection pressure on the biological aspect which evolves in this way over thousands of generations.

From those that led to the most fragile Australopithecus, an inconceivable break occurs, a break similar to that which led from physical chemistry to biology: that of the passage from the biological to the anthropological. Just as life emerged from matter, humans emerged from life. Life could have had other bases than what we know from physical chemistry, humans could have had other bases than the hominoid lineage; it turns out that in the contingency of evolutionary processes, this is the path that has been followed: starting from the great ape, a learning process that favors the sharing of attention, the pleasure of success, exchange between the sexes, an absence of sex selection, a group that is more in touch, and an interest in the tool.

Emerging technical capacities, and in particular bodily techniques, could lead to *Homo habilis* by various means that would otherwise seem unimaginable. In particular – whether it is the use of branches, leaves, bark, stems, animal skins and casings or bones – it seems highly likely that the latter developed long before the precise cutting of Acheulean stones. The know-how of stone knapping is much more difficult than that of branch cutting, although the latter will depend on sharp stone tools; on the other hand, the mastery of percussion, scraping, friction and fine cuts opens up infinite possibilities, and creates a selection pressure on the ability to make more precise tools. A chimpanzee can crack nuts by choosing the anvil and the hammer, which takes years. Why, once freed by the vertical position, are there no other possibilities? Let's bear in mind that archeological archives will have a hard time showing organic objects, even from the time of *Homo erectus*.

There should be an attempt to posit that certain techniques have been acquired and sometimes lost, from time to time reacquired, superimposed and associated. Today, we could make flints cut with perfect sharpness instead of knives, while using computers to write. But know-how, habit and our imagination prevent us from doing so. Before going any further, we should think about what can be achieved by groups of vertical pre-human beings who are already socialized, able to share tasks and able to leave "their fathers' environment", because now their world has grown and they are taking it with them.

There is no simple linearity in the evolution of know-how and knowledge. The understanding of what prevents or allows an invention or discovery, a development or a diversification, cannot be general. It must be looked at in context, and experience shows that even with this data, it is not obvious. Isn't this what stone knapping techniques show us? The discoveries of the Middle Stone Age in Africa which in the end lasted from 290,000 BP to 40,000 BP suggest that many techniques were put in place more than 100,000 years before what will be shown in Europe (Douze 2011). The fact that a technique is learned means that it can be lost, and the monitoring of techniques alone, especially after the emergence of *Homo sapiens*, will acquire a form of autonomy with respect to human biology which evolves much more slowly. It will have to be correlated with culture and needs, which can be associated with the environment.

Robert Bednarik (2012), along with Katja Douze (2011), argues that the so-called "modern human behavior" is much older than assumed. Everything seems to be more gradual, and again the beginnings are older than assumed. Neoteny begins with bipedalism, since human hands and feet are more like those of chimpanzees or embryonic bonobos than the hands and feet of adult monkeys. Would this Australopithecus foot have been obtained from neoteny? Isn't a selection via children's games an initial form of neotenization, in fact?

11.2. Phylogenesis of characteristics and lineages

It is a fact known to evolutionary theorists that selected traits may follow different trajectories from those of lineages conceived as more or less homogeneous populations, because individuals from different lineages may interbreed and give offspring that join another lineage or population. Thus the so-called "Eurasian" humanity contains genes from the Neanderthal and Denisovan Man, and if it is easy to distinguish at a glance the Far Eastern populations from the European or African populations, for the genes it is different. Almost all gene alleles[1] can be found in all populations; their frequencies can vary more or less strongly and the variability of the differences between groups is much less than general variability. What separates the groups visually is minor; what might separate them genetically is rarely noticeable. For this reason, geneticists believe that the term "races", as it relates to the human species, is not appropriate, even though it is possible to follow superficially certain marker alleles of group separation because they cause immediately visible shapes (skin color, hair structure, limb lengths, eye color, widths and shapes of openings, i.e. mouths, nostrils and eyes).

This data may allow us to understand the nature of evolution from Sahelanthropus, Australopithecus, Ardipithecus, Kenyanthropus to the genus *Homo*. Since we know that *Homo neanderthalensis* and *Homo sapiens* have known interbreeding, we can easily imagine that before *Homo habilis*, interbreeding was possible between all these supposedly pre-human forms. These are indeed the conclusions reached by scientists. Interbreeding between individuals of the lineage that leads to chimpanzees and those of the oldest pre-Australopithecus species took place. These interbreeds between slowly diverging groups may have involved species qualified as human (such as *Homo habilis*, which would be considered by many researchers as an "evolved" Australopithecus). These exchanges could have lasted four million years, the entire period from the beginning of the separation of Panina and Homininae to the emergence of *Homo habilis* (Patterson *et al.*

1 A gene can have several forms called allelic forms. The genetic heritage of a population is not only the sum of all genes, it is also the sum of all alleles. Genetic diseases most often come from an allele of a gene and not from the gene itself.

2006). The tree that leads to modern humans is actually a staggered tree that resembles a trellis where branches can coalesce and branch off again. The concept of species, which is already fuzzy in itself, seems to be even fuzzier here. We know of gradients in wide-ranging animals that make interferences difficult at the extreme edges and yet occur unproblematically in the intermediate regions (e.g. gulls).

The advanced occipital orifice (*Foramen magnum*), which corresponds to the head position on the body, moves, starting with Australopithecus; however, it is centered more with *Homo habilis* and increasingly so with *Homo erectus,* then *Homo sapiens*. This accentuated centration on the different species starting from *Homo habilis,* which takes place with an increase in cerebral capacity and decrease in the prognathism of the face, is a consequence: if the face diminishes and the back of the skull develops, the centration is accentuated. Thus, the advancement of the occipital orifice becomes linked to bipedalism, which exerts an incessant selection pressure; it is a strong indicator of the "lineage" of Hominінae.

According to current data, stronger knees with better resistance to standing appeared at least two and a half million years after the beginning of the Australopithecus; it still took 400,000 years for the foot to arch and the toes to shorten. This is a selection which followed the development of walking, or even an attempt at running; it is walking and running, which have become necessary due to the change in behavior, which in turn leads to a selection of the best walker-runners. This new ability comes from a "voluntary" walking system, which has already existed for almost three million years.

Canine teeth have become smaller, regardless of sex. In *Ardipithecus ramidus*, nearly six million years ago, the canines of both sexes became small and similar (Lovejoy 2009). This indicator of sex differentiation and polygyny – the rate of difference in canine teeth by sex – suggests that, for this species, males no longer fought to conquer the females; thus, it was long before the appearance of this tooth reshaping that the trend towards monogamy took hold and the males' fighting to conquer the females came to an end. What in a group could have prevented the selection of this ancient superior primate trait? We can imagine a process leading to early cooperation between the sexes in some groups at a much earlier time, when the bipedal lineages were differentiated, which corresponds to our hypothesis, because selection towards bipedalism seemed to be linked to that of monogamy. If it is similar in both sexes, as we have assumed, it reduced the differences between the sexes.

It was only 3.2 million years ago that the pelvis shortened and widened, that the thumb of the hand became elongated; it took 300,000 years, that is 2.9 million years ago, for the legs and waist to lengthen, and for the femoral joint to strengthen. Running, indicating active hunting, had become the selection factor: after the knees then the feet, hips and waist (lumbar) were modified. It is worth stressing that hunting cannot have occurred at that time. On the contrary, it is hunting that, having become necessary for the survival of the group, selected this evolution over time; the best hunters survived better (whether this selection is in a group or even inter-group does not change anything), because hunting was already indispensable for survival.

It can even be hypothesized that hunting was established, as a systematic behavior necessary for survival, at least five or six million years ago. Remember, everything took place long before it is visible in morphology; a morphological adjustment could only occur under specific selection pressure. The large number of modifications, over a relatively short period of time – as we would have to add the twisting of the humerus and the lowering of the shoulders (enabling ample arm movements, useful for hunting with tools) – suggests that all this may not have taken place at the same time in the same populations and may have been brought together by interbreeding.

Were all these characters selected in a precise order? On the contrary, archeological data shows that the different types found often mixed so-called archaic characters with other "advanced" characters (in relation to our representation of evolution towards becoming human) which became a puzzle for the theorists who wanted to achieve a linear evolution. On the contrary, this serves to emphasize the correctness of our hypothesis even more so. This conception of an orthogenetic, quasi-teleological evolution is yet another avatar of creationism that presents itself as Lamarckian.

While bipedalism is clearly based on the technicization of the body, each technical choice of a group creates a particular selection pressure that may require hundreds of generations and a specific progressive body modification. While bipedal primate populations spread over large areas, differentiated evolutions took place depending on the environment. The encounter of groups that were constantly on the move caused separately selected characteristics to be associated with each other that "form a system".

At the beginning of the 20th Century, the chimpanzee population was estimated to have exceeded two million. Today the 250,000 chimpanzees and bonobos seem to have a genetic variability greater than that of the 7.5 billion humans. Yet verticality favors nomadism and consequently interbreeding. In all the great apes, including the vertical ones, it is the females that leave the group to join another group; gene flows between groups are therefore important. The size of chimpanzee groups observed today is between 10 and 50, depending on food availability. They are now under great pressure from the expansion of humans, who are up to a thousand times more numerous around their territories.

It can be considered that during the great expansion of bipeds, six to seven million years ago, Panina and Hominінae together must have largely reached ten million over a total surface area of tens of millions of square kilometers. Africa was mainly green for a long time, even though there have been many climatic variations. Bipeds had to cover a larger surface area than other primates because of their wider adaptive capacities, as bipeds were better adapted to open spaces, and were probably distributed in hundreds of thousands of groups. It is now known from the study of gulls in the great Holarctic zone that distance and time eventually lead to different species, even if the overlapping areas allow hybridization. The same may have been true between bipedal species and also with less bipedal species.

This makes it possible to understand that particular characteristics may appear in different groups, as a consequence of different social practices and technicizations of the body, and that interbreeding makes it possible to associate traits that appeared in different populations. Even if the twisting of the humerus appeared at the same time as the lengthening of the legs, it is possible that this took place in different groups, and the meeting of these groups may have led to an evolutionary leap. It is functionality, embryology, and evo-devo genetics that make it possible to specify the order of appearance of the transformations.

We can conceive branch-like evolutions due to the expansion of bipeds, in many very different environments, with fusions of features favored by encounters of different groups after long journeys. This leads to a combination of incremental and more spectacular evolutions. Under these conditions, it is understandable that the places from which humanity originated (genus *Homo*) can be debated: near Chad? In South Africa? In East Africa? Or henceforth in Morocco? Since not all of Africa has been excavated – some specialists estimate that less than 3% of the potential surface area has been explored by researchers – it is highly likely that other sites will be found. It will also depend on the conditions that are specific to the possibility of fossilization.

Stabilized bipedalism, here posited as a result of a technicizations of the body, allowed humans to leave the forest, to overcome obstacles and to modify one's diet. As soon as the technicized biped, already Promethean, appeared, it moved more or less quickly, and discovered less frightening environments as the technical evolution progressed. Those that went too far (especially outside Africa) risked losing contact with other populations, their evolution through interbreeding slowed down and the environment was less favorable, ending in deadlock. Yet some went as far as China and Indonesia, like *Homo erectus*.

It is a constant of human evolution that any isolation of limited groups stops or slows down the evolution of that group. This observation confirms that human evolution is group-based and hierarchical. It allows us to understand that there have been several *exits from Africa*. It is indeed in the immense region that covers all of Africa, including the Sahara, which was green at certain times, with almost all the climates tolerable by a great bipedal ape, that evolution to human beings has continued. *Homo erectus,* the first one, seems to have shown the ability to go to colder regions; it has thus gone further. This is a question of technical level. Nevertheless, it seems that *Homo rudolfensis, ergaster* or *georgicus* or at least the archaic *Homo erectus* came out of Africa early.

Everything would depend on hybridizations that are as much biological as cultural; from *Australopithecus*, evolution is driven by culture and technicization, as the limiting factor to the selection pressure is biological, the evolutionary basis seems biological. We must understand that the foot ends up arching because any improvement in this area is preferred; walking and running is better, because the whole life of the hunting group demands it, not the reverse; the twisting of the humerus is selected, because hunting is what ensures survival; sweating and loss of fur are consequences, selected by long-term hunting and by the child protection that ensues. The use of tools and collective behavior create a selective environment, in itself, although the environment providing the materials for the tools and accessible food also has a selective role.

We have examples outside the human world of behavior that leads to partial biological modification and a form of physiological maladjustment. This is the case of the great panda, which is a bear, or a carnivore in phylogenetic terms. It possesses all the enzymes to digest meat and can digest only a small amount of cellulose. Its intestinal microbiota would be that of a carnivore. Yet its diet is vegetarian; its teeth have been partially adapted and it has a "false thumb" which allows it to peel bamboo, its favorite food. To date, the only reason for its lack of interest in meat is a mutation in a gene coding for a meat flavor receptor: in short, it no longer has a taste for meat. We can imagine that it is its interest in bamboo that selected this mutation or on the contrary that, protected in bamboo forests, this gene could have been selected because it led it to eat something else.

It remains that teeth and false thumb are characters selected after its interest in bamboo; on the other hand, its digestive system has barely evolved; it is a constraint that cannot evolve so quickly, so complex, which leads, in our "great panda", to long digestion periods. It is possible that it is easier to mutate from a herbivore digestive structure to a carnivore structure than the other way around. However, the wolf, domesticated as a dog, has undergone an evolution towards a more omnivore diet; the dog, after a few millennia of selection and consumption of starchy products, has amylases, whereas the wolf does not have them. Humans are ruthless breeders. The fact remains that bamboo is to the panda what the sea is to the dolphin and technique is to man.

As there are thousands of such examples in living beings, the first conclusion is that we do not know, *ex ante*, what could or may mutate. It is not yet known whether all mutations, without exception, are random, or whether some could be directed, almost pre-programmed, according to the environment. There are examples such as immune defense processes in mammals, or adaptation to these processes by a parasite such as *Plasmodium falciparum*; the biological system put in place makes it possible to manufacture randomly recombined components at the genetic level, which will be subjected to Darwinian-type selection.

In the living world, such systems exist. It is not known whether a species, under strong, directed selection pressure, could respond with more and varied mutations in the areas under pressure. This would seem Lamarckian and uncertain. As long as the variation-selection binomial is maintained, we are well within the probabilistic-selective approach. It is only afterwards that we see what has evolved, what evolution is built from (by variation selection), what has hardly changed and what does not change. Today, epigenetics suggests possibilities.

Why was the mammalian lineage that gave birth to cetaceans so well "adapted" morphologically speaking, while remaining homeothermic (maintaining internal temperature), breathing like a mammal, giving birth like a mammal, caring for the young even more than the majority of mammals? This is because some functions cannot be modified, even under strong and continuous selection pressure over a long period of time, as they are the result of complex and indispensable functions. It is easier to evolve towards a more important care of the young, characteristic of a mammal, than to transform lungs into gills or to give birth to young that are already adults. The structural and functional constraints specific to mammals cannot be modified over a short period of time, but it is the variabilities that are specific to mammals that will evolve. A group evolves in areas where it shows a high variability, provided that there is a selection.

11.3. From Australopithecus to the genus *Homo*: the selection of technicality

We still read that traits selected during human evolution facilitate the two trends in our evolution that are bipedalism and the use of tools (Wood 2014). Our theory here is that bipedalism, because it is a learned technique, and consequently also the use of the tools that are its correlate, are the selection conditions orienting this evolution. The driving force of the evolution towards the human would have started with the technicization of the body, which could only promote technical development. Each selection of a new genetic trait, under the pressure of the conditions created by technical use, fled to a higher technicization which in turn increased the selection pressure in the same direction. It is in this sense that I propose the Heideggerian expression "opening of the clearing". The technicized animal always asks for more technique, like the Goliath frog, which only seeks to become stronger than before. It is always looking for a vaster clearing; the opening of the clearing is a selective factor in the ability to live in this opening. It requires both genetic variability in the factor under consideration, a selective advantage over this variability, and selection pressure over time.

Thus, the first traits to evolve are all those that improve bipedalism, that is improve the technicality of walking. Just as a tool evolves, because there is variability in its realization if there is a pressure for it to be more efficient (the technicality of the human), so the body evolves by selection making it more efficient as a technical tool. This is how musculoskeletal innovations appeared, which became more and more usable for hunting. Each variant leads to collateral effects: the vertical position leads to an amplification of the relational, which itself favors collective actions, which are more efficient for a number of actions: hunting, cutting prey, intra-specific competition. This modifies the sexual relationship: the formation of the couple increased protection of children. The enlargement of the world is a consequence. This, in a technical society, leads to an exploration and increasing use environment; it therefore has a selective role.

The technique requires "double attention", concentration and cooperation. It calls for "joint attention", which means that two people are attentive towards the same object. It makes otherness appear, because this experience of the other is first lived internally: oneself, under the action of the body technician, becomes another: "I" is another. It is "the appeal to the consciousness", in the same sense as in the expression "knock-on effect". Everything "conspires" to favor mutations amplifying the consciousness of oneself, of the other and of the world. We can again take up the point of view of Robert Bednarik here. If the "theory of mind" defines the capacity of any animal to attribute mental states to itself and to others, and to understand that congeners have beliefs, desires and intentions, and that these may be different from its own, then the first superior primates in the process of bipedization had a theory of mind, admittedly weak, vacillating, variable and therefore selectable. There is no need for an explanatory theory about the theory of mind, consciousness, reflexive consciousness; it is enough to admit that they exist, in the sense that they are neuronally based and that there is some variability.

Fur loss and sweating are consequences of selecting traits to meet this need for exploration and broadening the horizon. We know that walking and running endurance are human characteristics; they are also consequences selected to meet this technical requirement for greater efficiency, this incessant tension resulting from verticalization. The increased efficiency of bodily techniques is associated with the increased efficiency of object techniques. This efficiency is not always rooted in neoteny, even though complete verticalization may, in part, have been at least partially rooted in the beginning. However, the increase in this efficiency, when technicization is present, is achieved through a decrease in "dentition, jaw, digestive system, direct body skills" specialization. The pressure of collective efficiency, in a changing and varying environment (climate, nomadism) effectively leads to a neotenization of characteristics. Thus, it is the slender Australopithecus that would have opened the evolutionary path leading to *Homo*. The more technical calls for the more technical by despecializing the use of the environment and the body. In technique, it is better to have a slender and skillful body than a more robust yet clumsy one.

We now read that man became a hunter more than two million years ago, earlier than we thought. Our hypothesis leads us to believe that we have to go back much earlier than that. Chimpanzees never turn their noses up at a bit of meat, but they have few technical means to access large or fast-moving species. A stone-throwing, stone-breaking vertical primate (pebble culture) can already increase its food spectrum. As access to a richer food selects the technicization of the body and the improvement of technical objects, technicalized hunting obviously started long before the admirable adaptation to hunting. If, more than two million years ago, our ancestors were already good hunters, as the latest discoveries suggest, this means that hunting began long before. The hunter does not have to be a champion; it is sufficient for him to be "good enough". Just as the mother who is "good enough" according to Winnicott (2005), to ensure our existence, our ancestors in the verticalization process became "good enough" game hunters.

It was the Australopithecus groups (or their ancestors or related species), inventors of hunting, who created the conditions for human evolution. The others eventually became extinct. Hunting, although it provided better nutrition, did not prevent us from using plant resources; the first hunters were, biologically speaking, more vegetarian; becoming omnivorous made us stronger in the competition for plant-based food. The collective organization for efficient hunting requires additional energy, which is rewarded by a richer diet. To confirm our point, a publication gives "evidence" that ancient Australopithecus more than 3 million years old were able to cut, scrape, grind bones to extract meat and marrow (McPherron *et al.* 2010). Even more recently, discoveries in 2012–2015 at the Lomekwi 3 site near Lake Turkana in northern Kenya have revealed cut stone objects dating back 3.3 million years and attributable to a species of the genus *Australopithecus* or *Kenyanthropus* (Harmand *et al.* 2015). In this context, they were already hunting.

The argument is that any living organism that is well adapted to a given activity or a given "niche" is the descendant of populations that have undergone stable selections for a longer time; but its activity creates new selection conditions. Since human evolution is driven by technique – in this case hunting – it means that humans began hunting long before they became the world champion in any category. But let's not get bogged down with hunting: having sharp stones can help you to cut branches (habitat), open various fruits, cut skins, guts (for clothes, shelters, ropes, etc.). There are many other things you can do when you are able to knap stones.

How has our big brain been selected? It was also by technique, once the technique had allowed the selection of the physiological changes that made it possible. In order for the skull to develop, the maxillary muscles had to be diminished. It would have been about 2.4 million years ago that the monogenic mutation, reducing the size of the maxillary muscles, would have appeared (Stedman *et al.* 2004), several hundred thousand years after access to rich food no longer requiring powerful jaw muscles. But in return, this shortening requires the maintenance of a rich diet.

How could this change, which made chewing twenty to fifty times less efficient, have established itself? The answer is that powerful chewing was no longer useful; on the contrary, it expended a lot of energy. Access to rich animal feed no longer required powerful grinding, and stone tools allowed for food preparation. Any change in this area was no longer counter-selected. On the contrary, it opened up other advantageous possibilities. The energy saving resulting from the disappearance of these powerful muscles, and the time saved in accessing a richer diet, throughout the year, made the group more efficient. The technique for weakening the digestive system existed long before this biological transformation. However, this biological transformation led to irreversibility; naturalization no longer became possible, and the technique became indispensable. We can notice that the widening of the hips, as a consequence of the verticality, may have resulted in a baby with a bigger head being possible. This adaptation of the hips to the vertical posture is indeed prior to the appearance of *Homo habilis*; it is the result of an evolution towards a stronger verticality. Compared to chimpanzees, Australopithecus have, in relative terms, narrower shoulders and wider hips.

To take the variety of Africa's climates in space and time, between seven and two and a half million years BP, as a driving force in itself for pre-human evolution is a bit exaggerated, although it carries a significant element of truth. Why would this process have worked on hominid lineages, but not on those of lions, giraffes or elephants? It is indeed the technicization of human groups that has been favored by spatiotemporal climate variability, because it accentuated the advantage of access to a varied diet and adaptation (Lamarckian, for once) to climate fluctuations. When climatic conditions changed, the more technologically advanced adapted; the more naturalized could not resist and disappeared (de Menocal 2011). Climatic oscillations acted as an accelerator because the technicity already existed. The fact remains that during the evolution of hominids, there was also the evolutionary waltz of felines, antelopes, proboscideans, all changing but remaining more or less in the niches of their ancestors.

This we know: the most technical human societies adapt more quickly and are more combat-oriented. The bodily technique is first of all a fight against our own body; it is an incessant type of training and a driving force. It leads to the development of external tools which, through feedback, create the conditions for the selection of an additional technical approach. In this sense, the spatiotemporal climatic variations in Africa during this period have been able to accelerate the evolutionary process. In any case, they were a factor that favored the most adaptable groups, those which, in this particular case, were the most technical. Even if we cannot eliminate *a priori* the simple genetic drifts of isolated groups, if the technique does not bring advantages, we do not see how it can be maintained. If it is maintained, it inevitably leads to biological evolution in the long term.

Technicization with tools, because it requires a technicization of the body, is a struggle against the body itself. In a metastable system, this imbalance is a selection pressure. The ancestors of humans created a powerful engine of evolution by being in a permanent state of imbalance. Since the beginning, technique looks to be a form of violence. The body's response, in the succession of generations and selections, is to become more flexible, within the limits of its possibilities. The body yields to technical pressure by responding to it in two ways: internal technicization (a kind of reactive passivity and flexibility) or fragility (giving tools the ability to act and mediate with the environment). The brain develops accordingly.

The most technical group may not be the most important: it only needs to be able to develop. It will be able to explore more varied contexts, build its own environment and broaden its horizon. During a climate transformation, naturalized specialists can be swept away and the technicalized groups can take their place. There can be interbreeding and combinations, both sociotechnical and techno-biological.

12

Technical Evolution and Neoteny of the Genus *Homo*

We have analyzed the neotenization that led to hominization, that is to the great upright apes, under the selection pressure of a new environment combining bodily technicality and equipped technicality. The Lomekwi site became an indicator that the production of stone tools did indeed come from these Homininae. The knapping of stones appeared before the growth of the brain. *Homo habilis*, long considered as the emblem of the passage to a new cycle, that of "humans", changed status. Wouldn't it simply be an indicator of the end of the reign of the pioneers of verticalization (Orrorin, *Sahelanthropus*, *Ardipithecus*, *Australopithecus*, *Paranthropus*)? Some believe that *Homo habilis* was just a form of *Australopithecus* with a larger brain. The debate is likely to remain open for a long time, because the divisive lines in an evolutionary process are never clear-cut. Between 2.5 million and 1.5 million years BP at least seven "species" of Homininae were identified. In particular, *Homo habilis* and *Homo erectus* probably had several hundred thousand years of living alongside one another, in slightly different ecological niches. We find a situation similar to that of the early Homininae: a bushy development (Spoor *et al.* 1994; Leakey *et al.* 2012; Spoor *et al.* 2015).

12.1. *Homo habilis*: a new bushy development?

Homo habilis had a brain volume between 550 and 680 cm^3. In terms of weight, its cephalization index exceeded that of the great apes, including *Australopithecus*, without being excessive. The birth of a baby with a larger head was made possible by the vertical posture, which over thousands of centuries of selection has led to a significant shortening and widening of the pelvis.

The vertical position required new qualities, while the effort to control this verticality required more neurons and connections between perception and action. In fact, this verticality led to looking into the distance; a double tension of elevation, distancing, and gaze that pushed back the horizon. Moving from an approximate verticality like that of an *Ardipithecus* or an archaic pre-*Australopithecus*, still limited to the forest, even if sparse, which closed its potential horizon, to that of a *Homo habilis* who had access to all landscapes and the immensity of the savannah, required musculoskeletal transformations and considerable neurological development. The synchronized re-entry of more and more neurons, indicative of consciousness, would have already been increased; a form of expanded consciousness was being put in place. The improved technicality selected specific cerebral developments; it selected the multifunctionality of many cerebral areas. When a new technique appeared, the response to the selection was both the appearance of new brain lobes, the development of old ones, and the interchangeability of neuronal functions.

It is known that certain neurons seem to be involved in functions that are related to so-called higher mental abilities, in particular: the theory of mind, consciousness, reflexive consciousness or self-consciousness, but also recognition in front of a mirror and empathy. These are spindle neurons or von Economo neurons, or VEN, named after the person who discovered them in 1929. There are just a few tens of millions of them concentrated in the anterior cingulate cortex, the fronto-insular cortex and the dorsolateral prefrontal cortex of the brain of hominids, including great apes. These spindle neurons have been found in dolphins, whales and elephants. The technique, therefore, cannot be the direct cause of consciousness. Nevertheless, it is indeed these brain regions that increased in the passage from *Australopithecus* and fellow creatures to the genus *Homo*.

This was a real socio-technical and techno-biological transformation. It is understandable that a little more than four million years were needed, four times the average time it takes for a new species to emerge. Many naturalization pathways, that is ultimately specialization, developed and eventually failed in the context of Africa's high spatial and temporal variability, between 7 and 2 million BP. New species would be continually found on the pathways of eventual hominization, bearing in mind that there was only one succession that led to *Homo sapiens*, but interbreeding complicated the evolutionary tree. The effect of the double technicization of the body and tools demonstrated its strength; something new emerged on the part of the body. The brain is integrated into this bodily transformation. The reading of paleoanthropologists' descriptions of *Homo habilis* reveals the profound reasons that led those who discovered it to now define the genus *Homo*. We are no longer in the beginnings of humankind, as primatologists studying the great apes of the present time are discovering; we are indeed in a world that qualifies as human, even though *Homo habilis* may not be considered so much *Homo*.

This was a world of groups of up to 80 individuals, according to Dunbar (2007), specialized within each group, that is socialized, able to go out in teams to find the food that suited them. They were omnivorous even though plant food remained dominant. The individuals in the groups helped each other. They started to find out how to exchange information and achieve technical learning. Walking became assured; the selective pressure of verticality favored it; brisk walking and running were flourishing, even if changes – described as improvements – took place afterwards. They eventually exited Africa and spread far and wide (might *Homo rudolfensis*, *ergaster*, *georgicus* be indications of that?). They still lacked, given our position as modern humans, a superior capacity for conceptualization. *Homo habilis* would exist today to the extent that we would be ready to consider them as brothers, even more than the chimpanzees and bonobos; we would vote laws to defend them. But their ability to exchange would remain limited.

Contact with bonobos or chimpanzees is already disturbing. The acuity of the gaze, the moral and collective sense, the emotional and affective capacity, all this is so familiar. A chimpanzee dancing in the rain, trying to vocalize, expressing jealousy, confuses the most rational primatologists. *Homo habilis* added something that seemed unlikely: the capacity for bodily technicization, precision of gesture, relational closeness, the ability for a differentiated relationship, and probably the nuclear family. It is probably around this time that the sclera of the eye became white, which can be confirmed by molecular biology; consequently, following the gaze of the other, indispensable for collective life, became easier. Of course, it would remain difficult, if not impossible, to have an elaborate conversation with him/her, but we can imagine that a collaborative work activity would become possible, because the technical activity of *Homo habilis* was learned and transmitted: he could have lost all this knowledge accumulated throughout millennia; in this case, the group concerned came to disappear. But this also meant that from that period onwards, techniques evolved more independently, with interspecies transfers. As the saying goes, any technique is learned; if the body is adapted to it, this learning could come from anywhere, and the loss could take place as soon as the use disappeared.

With *Homo habilis*, technical development, built on the technicization of the body, was approaching a breaking point. The chopper was invented a long time ago; with the "Lomekwian" culture (3.3 million BP) then Oldowayan one (2.6 million BP); the knapped two-sided stone culture is almost there (Acheulean). Everything that would allow thinking seemed to hatch; the basis of what would allow thinking and speaking about thinking was present. All that was missing was a new, specific tool which, like a mirror, would return the technical unthought – unthought because it was the foundation of the possibility of thinking – to the technical actor himself, enabling him to transmit his technicization in usable terms. But was there a need for this? In *Homo habilis,* through direct learning, mimicry and trial and error, transmission took place in a context of technical tension and functional communication. Double attention and joint attention were already present, as well as gestures associated with sounds and sound articulations: a kind of pre-language. Following a theoretical approach, it seems coherent that *Homo habilis* used a pre-language made of sounds and gestures and, who knows, maybe even whistling (Meyer 2008).

This new tool was immeasurable when compared to existing tools. It was in a way the "tool of all tools" the reflexivity of tool use. It was no longer a musculoskeletal issue, nor even a simple question of social organization and task sharing or collective coordination. Wolves and chimpanzees are capable of cooperative collective organization. But they lack the interaction of these collective and technical capacities in order to arrive at the human, in the framework of a reflexivity about what is happening in order to rebuild and exchange it. In this respect, *Homo habilis* seemed almost as powerless as the chimpanzee or the wolf. At most, there was a pre-language, because Broca's area, already important in the great apes, is more developed. Note that some dolphins seem to have some sort of pre-language. Could the whistling be a prelude or is it a divergence?

Yet a technological leap was about to appear, the one that separates *Homo habilis* from *Homo erectus* and *Homo heidelbergensis*. If the first neotenization concerns firstly the posture, the second concerns firstly the brain. About 2.4 million years ago, according to molecular analyses, monogenic mutation, which reduced the size of the maxillary muscles, appeared (Stedman *et al.* 2004), after several hundred thousand years of access to a rich food supply, hunting and – who knows? – cutting, grinding, soaking and fermenting food which was richer, more digestible and required less and less powerful jaws. The comparison with the dating of paleoanthropologists is troubling. It would have taken place on a contemporary lineage, but different from *Paranthropus boisei* and *robustus* and also from *Homo habilis* since the latter was present at that date and continued to exist for at least 600,000 years.

While *Homo habilis* still lived their quiet lives, it was in the so-called *Homo erectus* that the maxillary musculature and the modern dentition, the modern digestive system (not directly detectable, but logically deduced), the modern femur and femoral joint, the "barrel-shaped" rib cage, the flexible spine at the lumbar level, and ... a brain that reached or even exceeded 1,000 cm^3, appeared almost concomitantly. At that time of the end of the *Homo habilis* lineage, there were many types or species of related humans. The branch-like evolution was there, evolutionary barriers were overcome, after an accumulation of mutations selected by the pressure of technical tension or social organization, or other uncertain but associated parameters such as climatic disruptions.

It seems coherent, to date, that in terms of filiation, *Homo heidelbergensis* can be considered a link that led to *Homo sapiens* and *Homo neanderthalensis*. However, they cannot be on the same genealogical lineage if we compare dates and differences between them. For simplicity of reasoning, we will keep only three milestones: *Homo habilis*, *Homo erectus* and *Homo sapiens*. Another genealogy, based on the genetic analysis of human lice (head and body) has been proposed by Reed *et al.* (2004) and does not change our thinking. The filiation would be as follows: *Homo ergaster* from which *Homo erectus* and *Homo antecessor* came; from the latter came *Homo rhodesiensis* from which *Homo neanderthalensis* and *Homo sapiens* came separately. Fertile interbreeding was possible, and certainly took place, which allows us to understand that "archaism of characters" could be varied according to individuals or groups.

Our hypothesis here is that the configuration prior to *Homo habilis*, around 2.6 million BP, became “metastable”, as well as that of the beginning of verticalization (*Ardipitechus et al.*). *Homo habilis* was end of a lineage not leading to humans. When we find many skeletons, we are at the maximum of quantitative development. New forms came from small populations. Around 2.4 million BP, thanks to technical innovations opening up new biological possibilities, new species appeared. A new branch appeared, after the one that produced the *Australopithecus*: the protohumans. How can we cross the threshold, which can only be conceived after the fact, that is to say, in the historical knowledge of what has taken place, other than through a co-development of increased technicizations of the body and more advanced collective techniques? Everything is there to allow it, but the brain is insufficient. Here again, selection promoted a new heterochrony associated with a morphological regression of a whole: the digestive system. This regression was made possible by the available technique: a now rich and varied diet, global isolation of the group which enlarged the clearing of the being, to use the Heideggerian expression, the possibility that childhood was longer and that birth was premature. Let us analyze these different traits.

12.2. *Homo erectus*, the advent of a technical humanity

It would appear that *Homo erectus* had a broadly similar digestive system to ours. The jaw muscles were shortened, they no longer attached to the top of the brain; as a result, the pressure on the skull decreased, the brain could develop without antagonism, by simple heterochronies obtained by a modification/mutation of so-called homeotic regulation genes in the development of the brain. The face remained that of a primate fetus, with a diminished jaw and weaker teeth without the need for long chewing time, according to the same type of event; the digestive system could be reduced without risk, since food was rich and digestible, and it was easier to diminish an organ than to make it more complex. Here we see that the transition to *Homo erectus* corresponded to a “massive” neotenization. All of these modifications promoted a vertical posture that was even more agile in the use of tools. Each modification could be independently promoted, it appeared in different groups and combined with others by interbreeding.

Paleontological data indicates that this major transformation took place between 2 and 1.5 million BP. *It is striking that the decrease of the maxillo-digestive system appeared, genetically, a few thousand centuries earlier*; it was selected by the fact that it improved agility and body flexibility, under the selective pressure of technical verticalization ... but it took time. A lighter stomach allowed for the release of the flexibility of the lumbar vertebrae, increasing the possibility of long walks or even endurance races. It also corresponded to the consolidation of the shoulders and arms, as well as to the development of the lungs, even if they could be obtained separately and recombined with the reduction of the digestive system. This development corresponded to the current human situation; for long, sustained running, the upper body muscles needed to be well developed. This is what *Homo erectus* was aiming for.

How could this transformation, making chewing 20 to 50 times less efficient, be necessary? The answer is that powerful chewing was no longer useful: it expended a lot of energy. Access to rich animal feed no longer required powerful grinding, while stone tools meant that food could be prepared. Vegetable products were fermented, removing the anti-digestive factors. Any change in this direction was no longer counter-selected and resulted in other advantageous possibilities. The energy saving resulting from the disappearance of these powerful muscles and the time saved in accessing a richer diet, throughout the year, made the group more efficient. The technique, both hunting and food processing, allowing the digestive system to weaken existed long before this biological transformation. In fact, there are more and more publications on the diet of Homininae before *Homo erectus*. We still have a lot to discover. Certainly, the invention of fire was a breakthrough, but before that the food was already processed and fermented, which improved the digestibility of food (Carmody and Wrangham 2009; Wrangham and Carmody 2010; Henry *et al.* 2014; Zink *et al.* 2014; Zink and Lieberman 2016; Wrangham 2017).

This transformation was a sign of the appearance of *Homo erectus*, carrier of the gene decreasing the size of the maxillary muscles, since its brain reached 1,000 cm^3 and its dentition was strongly diminished. According to Dunbar's (1992) curve, *Homo erectus* lived in groups of about one hundred individuals. *Homo erectus* and *Homo habilis* could not be on the same filiation. They differed on three essential points: sexual dimorphism was more pronounced in *Homo habilis* than in *Homo erectus*, diet was more carnivorous in *Homo erectus* and its nomadism accentuated.

For less than 500,000 years of initial difference between the two species, there were too many morphological differences between *Homo habilis* and *Homo erectus*; moreover, they were contemporaries and their ecological niche was different. A reasonable hypothesis would be that *Homo erectus* came from an unidentified species, a contemporary branch (competitor?) of the beginning of *Homo habilis*, already more carnivorous (more hunting-inclined) and more technical; its appearance must have taken place after 2.4 million BP, because having a shortening of the maxillary muscles and having at the same time a diminished jaw and dentition and an increased brain cannot be simultaneous; it was a consequence in the duration of a pre-individual technical environment in co-evolution. The earliest date of recognition of *Homo erectus* is 1.9 million BP, that is 500,000 years after this mutation of the maxillary muscles; if we are lucky, paleontology will allow us to go back more than 300,000 years into the past by finding the intermediate forms. The fact that no close ancestor of *Homo erectus* has yet been found suggests that the populations were small, or that it appeared in an as yet unexplored part of Africa. It is credible that transformations concerning both brain size and a dentition adapted to a rich diet were possible with few mutations. However, this posed other problems. Certainly, *Homo erectus* was more upright, more of a hunter, hips wider and a baby with a bigger head was possible, but how was it possible that dentition was already adapted to less hard and richer food? This decrease in molar size was too rapid to be a consequence. A technical selection was needed to both accept a baby with a big brain and to offer a rich and easy-to-assimilate food.

All the selective conditions had to be in place more than two million years ago. What could have happened, technically speaking, to select the spread of these major transformations? Timothy Taylor (2010) assumes that it was during this time before *Homo erectus,* that is in an *Australopithecus,* that the "baby carrier" or sling was invented. When we see what a chimpanzee is capable of doing with stems and leaves, stones, or different objects found in nature, we do not see why a contemporary group of *Homo habilis* – a better hunter than it and more carnivorous, capable of breaking stones to get a sharp edge, scratching bones and skins and splitting bones – would not be able to make a baby carrier out of skins and guts. The invention of the baby carrier had a huge advantage: the mother was less vulnerable, she could remain active and she did not slow down the nomadic group's movement. The baby carrier was a major asset to allow the movement of a hunter-gatherer group. It had another advantage: with a baby carrier, the mother was in less of a hurry for the child to become independent. This meant that if a mutation of a regulatory gene led, by heterochrony, to a mutant which had a larger brain or was born a little prematurely, having a baby carrier eliminated any counter-selection as the child could be carried for a few years; births were sufficiently far apart.

This larger head was now made possible by the removal of the maxillary muscles, as we saw earlier. While the rich and "processed" food enabled the selection of a reduced digestive masticatory system under the pressure of the selection of an ever more efficient verticality, the baby carrier eliminated the counter-selection of a larger brain and a late maturity. The baby unable to walk or cling to its mother appeared; its necessary learning of the upright position was even more visible, but, as it had been in existence for a long time and had selected a deep neurological mark, its progressive lack of precocity was still acceptable. It's important to understand that while the baby carrier was invented, all the means to protect the little one, to wrap it, to wash it, were also put or were being put in place. This was possible, because correlatively they were more relational, more cooperative, more enterprising humans. It is conceivable that during the time of the invention of the baby carrier, humans had lost their fur at least in part; the baby had nothing left to hold on to, but knowing how to hold on to became useless because it was wrapped up and carried. Brain development, obtained by a few mutations already present and until then counter-selected, flourished. Timothy Taylor argues, in line with our approach, that it is the technique that led to this situation: we would be "artificial apes".

In addition, a female *Homo erectus* pelvis, dated approximately 1.2 million years ago, suggests that *Homo erectus* could give birth to babies with larger heads (Simpson *et al.* 2008). The female pelvis of *Homo erectus,* at that date, was about the same width as that of *Homo sapiens,* and the channel through which the baby passed was about the same diameter. This was the result of increased verticalization. There was no obstacle to the birth of a baby with a larger head.

We can then ask the question of processed food: what are we speaking about here? On this subject, we can be imaginative. First of all, it is now known that chimpanzees are not especially afraid of fire. They observe it, follow it and do not panic. In a way, they conceptualize it and observe its effects. Sue Savage-Rumbaugh, a primatologist at the Great Ape Trust in the town of Des Moines, Iowa, studied a bonobo named Kanzi who spent his childhood in contact with humans; at age 31, Kanzi developed a primitive sense of cooking. Kanzi had seen many movies, including *La guerre du feu,* which he had seen many times. He had "long been fascinated by the way the guards in his camp cooked food and was encouraged to interact with humans and copy them". On a video, available on the Internet, he can be seen making a fire, gathering wood and dead leaves, while taking into account the position of the wind. Then he grabs a frying pan, puts it on a grill and lets the contents simmer. In another, he prepares a fire to roast marshmallow, a delicacy for him[1]. However, bonobos are not known for their use of tools in nature.

The conclusion is thus quite easy. It seems obvious, on the basis of the morphology and physiology of *Homo erectus,* that fire use came from very far away. We can make the hypothesis here, consistent with our analysis from the beginning, that the use of fire and/or fermentation had a very important role in the selection of the regression of the jaw and digestive system of *Homo erectus*.

This is suggested by a recent publication (Organ *et al.* 2011). The conclusion is that the cooked diet appeared earlier than estimated. The possibility of cooking, even in the "archaic" way, would help us to understand how *Homo erectus* appeared. The absence of fire, or the extreme difficulty of detecting it, is due to the nomadic lifestyle; but as the authors of the article show, the analysis of the evolution of the anatomy leads to the conclusion. That said, it seems clear that the use of fire is much older than is officially claimed today (Gowlett *et al.* 1981; Karkanas *et al.* 2007; Beaumont 2011; Hlubik 2018). *Homo erectus* did not invent fire: their existence came from it.

We can also go further and posit with Sonia Ragir (2000) that we need to reconsider the diet of pre-humans. It had to be processed as soon as the ability to make stone tools was proven, that is cut, crushed, fermented using a variety of foods of animal or vegetable origin. Whether animal proteins, tubers, rhizomes or bulbous plants, as soon as the processes enabled their digestibility, it was a new form of access to food that widened "human clearing" as Heidegger said.

1 This film is directly accessible at the following address: http://www.maxisciences.com/singe/un-bonobo-allume-un-feu-de-bois-pour-faire-griller-des-guimauves_art32300.html. Copyright © Gentside Découverte.

The possibility of food preparation and consumption, even in an "archaic" way, that is based on fermentation rather than cooking, would make it possible to understand how *Homo erectus* was selected. Bipedalism, stone-knapping, hunting, cooking, the domestication of fire, the protection of children, etc. all began before the supposed dates, on earlier lineages in terms of dates at the end of *Homo habilis*. The latter was an end of phylum, not the ancestor of *Homo erectus*. The domestication of fire was an event in the evolution towards humankind as it was the bearer of a new world (Bachelard 1949). We can say that it was much earlier than what is affirmed, and therefore more structuring in the human unconscious.

Let us add that the domestication of fire had as probable corollary the selection of the control of the breath, because to maintain a fire, we know instinctively that it is necessary to know how to blow in a controlled way. However, breath control is indispensable for fluid speech. With the mastery of hunting, a technique for the protection of children, the interest in cooking, the domesticated fire and a protolanguage that allowed the transmission of know-how, *Homo erectus* was completely out of the "normal" animal world. He tried to conquer the world. He went as far as China and Indonesia.

While the population was small, the conditions for constitution (the study of which is called taphonomy) and finding paleontological records are such that they are unlikely to indicate to us when fire was used routinely, especially if the group was nomadic. As fire is made by rubbing wood, that leaves nothing either. It can be noted that knowing how to make fire is easier than knowing how to carve stones. You just have to try it yourself. The initial domestication of fire could therefore date back about 2.4 million years. All we have to do is make discoveries to find traces of it.

The domestication of fire has had an enormous impact on the evolution of Homininae groups, and this in several different ways: protection against predators, richer and healthier food, breath control, predictive and collective management of time (because a fire needs to be fed continuously), the creation of collective "evenings", with a development of a sense of more personal relationships. Here we join Robert Bednarik (2013): we can claim that the beginning of humanity, even if it did not have the brain that it would have later (and that we no longer have), began long before *Homo sapiens*, in the lineage that produced *Homo erectus*.

Anyone who has tried to make a fire at a young age discovers that the use of flint, from "stones to fire", is easier than the production of bifaces. Stone carving requires a completely different dexterity than starting a fire with a stone. The conclusion is evident; once the interest of producing fire using a naturally occurring fire is discovered, the individual who knows how to carve stones will have the experience producing sparks and recognizing the smell of fire; he will seek to make fire. He will then be able to find other solutions. Fire is protection for a group and makes the night safe. It is likely that the use of fire first had a protective purpose and then its use spread to harden the tips. And what could be easier than putting a tuber in burning ashes? The absence of any trace of stable fire before 800,000 BP is linked to the behavior of nomads who avoided leaving traces. No nomadic fire, well cleaned when leaving, leaves a trace after thousands of years. This does not prevent prehistorians such as Henry de Lumley (2017) from estimating that fire was domesticated only 400,000 years ago! He confuses evidence of sedentary life with evidence of fire domestication. In fact, he takes up the logic of the hunter man considered as such when he became universal planetary champion. He had begun millions of years before.

Cooking and the baby carrier, the use of hides as protection, well-developed hunting and gathering tools, the ability to build shelters, among others and in any case associated, would have built the conditions for the emergence of *Homo erectus*. Indeed, when one knows how to make fire, break stones to make scrapers or percussion tools and knot a baby carrier, it is obvious that one knows how to do many other things. We can add: increasingly precise throwing weapons, sharper spears, spikes, stakes, an increasingly elaborate collective hunting strategy, increasingly protective habitats, clothing made of skins, and an increased interest in protected and surrounded infants at birth. *Homo erectus* was the result of a neotenization made possible by the development of technique at two levels: bodily technique, because the baby, even a neotenic baby, was subject to a requirement of continuous verticalization, to an increasing number of body and behavioral apprenticeships, and to a technique that was increasingly well equipped.

Homo erectus came out of Africa. On the one hand, he went as far as China and on the other hand he was found on Indonesian islands. He was able to adapt to the cold, build boats and sail. If fire – and the baby carrier – was used before *Homo erectus*, it concerns the oldest proof of navigation (Bednarik 2003). Claiming that he could not speak seems inconsistent with the recognition of such abilities. But was it a protolanguage? And the question remains: why did he not continue? The answer lies in the workings of evolution. Only one lineage has remained the same. Others gave other possibilities, one of which is well known.

12.3. *Homo sapiens*, the advent of inner life and the imaginary

Now it is a question of understanding the passage of a neighboring lineage of *Homo erectus*, which could have led to *Homo heidelbergensis,* a possible candidate of an intermediate link leading, among others, to *Homo sapiens*, the man called, or allegedly called, “anatomically modern”. Whatever the supposed date of the actual use of fire, archeological evidence confirms that *Homo erectus* used fire in a stable and continuous manner, in a sedentary lifestyle, at least 300,000 years before the emergence of *Homo sapiens*. We have understood what this early date means. Similarly, paleontological records show that, at that time, the corresponding species possessed the neurological and anatomical equipment necessary for the production of speech; it is evident that it had long since mastered breathing. While everything seemed to be in place for speech, for individuals of the *erectus* type, hundreds of thousands of years before *Homo sapiens*, this did not mean that these humans spoke as we do: on the one hand, neurological organization needed to allow it, and on the other hand, the selection of such an aptitude must have been present for many, many generations. The brain of *Homo erectus* would seem insufficient to produce complex representations, but this is debatable, and should be left to cognitive scientists and neurologists to study. The hypothesis here is that *Homo erectus* was a fully-fledged human being. Moreover, it shows such morphological variability over time that it is possible to understand the quantity of species described by paleoanthropologists over this period of time. We are here at the third bushy branch in the evolutionary development towards humans, one branch of which led to *Homo sapiens*.

We have many arguments for believing that unfinished speech was functional at that time. It is through speech, as we know it, that the field of possibility was widened and that humans could transmit both knowledge and know-how that was properly technical; it was through this that possibilities of objectification and subjectification could be established. Technical evolution during *Homo erectus*, and during the earliest Paleolithic period of

Homo sapiens, suggests that language, in the finished form with which we are familiar, was not complete and we will study it below.

Understanding the *Homo erectus* phenomenon is the key to understanding the next step. This species lived for more than 1.8 million years, a record in the life span of Homininae species, perhaps shared by one or two *Australopithecus* species. How was it possible? What does this tell us? It shows a stable state and adaptation to environmental and climatic upheavals, as *Homo erectus* encountered them. With the use of fire, technical child protection, a proven hunting method, stabilized stone-knapping technique (Acheulean), efficient woodworking and effective protolanguage, *Homo erectus* was prepared for a long nomadic life around the world.

However, with the emergence of *Homo erectus*, other possibilities developed that maintained a metastable state. This led to both diachronic and synchronic multiplications of species, weaker, more fragmented populations and smaller archeological traces. *Homo erectus* lived through a million years, during which new species whose future performance was shown, emerging from the *erectus* pattern and acquiring an even larger brain while practicing more efficient stone knapping known as the "Levallois method" (Boëda 1994, 2013; Eren & Lycett 2012). Why, and more importantly, how?

We get the following approximate sequence: pre-language with the lineage leading to *Homo habilis,* because the vertical position freed the larynx; brain development according to the modalities already described previously, because the pre-language, as a tool, created new conditions of selection; possible morphological alterations improving the possibilities of vocalization already present; cerebral differentiation selected by the impact of a more and more nuanced vocalization; emergence of a protolanguage, around *Homo erectus;* and emergence of a double articulated speech allowing the representation of reality, which, associated with the ability of imagination, leads to the invention of the human individual in the "modern" sense. *Homo sapiens*, the "morphologically modern" man, would appear on a blurred boundary between the last two stages; the last stage being neuropsychosocial, that is not biologically detectable. And before that would have been *Homo heidelbergensis*.

For Stanislas Dehaene (2009), "our brain is not a clean slate of cultural constructs, it is a highly structured organ that makes new out of old. To learn new skills, we recycle our old primate brain circuits – as long as they tolerate a minimum of change”. From the emergence of speech, this would become the rule, which makes it possible to understand the amount of future innovation and the adaptation of the human brain to the new tools that would be developed. This process corresponds to the concept of exaptation which will be taken up and developed further.

We have shown on several occasions that any proven date could imply a much earlier one; here it seems almost obvious that it was so much easier to make fire with flints than to carve a biface that any young person who tried it could see it. Bachelard (1949) has described the prodigious imagination provided by the presence of fire, the same fire that the Greek myths imagined was offered by Prometheus, proof if any were to be found of the importance that was given to it. We are entitled to estimate that, between the beginning of the domestication of fire and the emergence of *Homo sapiens*, more than a million years have elapsed, more than enough time to create a new species through artificial selection, based on new conditions of existence offered by “modern” technical activity: fire, cooking, clothing, habitats, throwing weapons, cutting capacity, etc. There is no shortage of food, time spent making tools, creating clothes or searching for the necessary materials, and learning is increasingly important.

Community life functioned around a fire; child protection systems, including the baby carrier, and an increasingly specialized social organization made life easy, to the point where the shifts followed one from one another. *Homo erectus* had set out to conquer the world, as far away as China and the Indonesian islands. He could face the cold, the sea and meet new worlds. This meant clothing, protection for the feet, careful protection at birth, in short, everything necessary for increased neotenization under continuous technical and cultural pressure ... to leave, to visit the world, to discover new territories. The capacity of tonalization and speech was still primitive, in any case it was a question of exchanges about goals, means, interests, affective exchanges; the double articulated representative language was not complete, but what means did we have to prove it? What if whistling had been the beginning (Meyer 2008)?

Chimpanzee vocalizations are known; it can be estimated that under the continuous pressure of technicization, the need for transmission has increased (see Pradines, Sigaut, Sloterdijk, Tinland). It is difficult to know whether a language with hands was first developed. It is a hypothesis that it is possible to hold (the cerebral areas of language and those corresponding to the sensorimotor aspect of the hands are very close, and today so many humans need hands to speak, especially since speech is conceptual). The chimpanzee cannot provide us with any information on this subject, precisely because its manual control is lower compared to that of humans.

This is another logic of language selection. The technique requires precision; the articulated parts of each technical or technologically derived object, their assembly and the finished object must be named. Attention to the object, which is specific to the technicality and its transmission, requires naming. The technique, as soon as it develops, requires a growing vocabulary.

For a language to emerge in such a way that multiple languages are imaginable, many morphological, physiological, muscular, neurological conditions are required, and most probably different selection pressures depending on the order in which the different modifications may have taken place; not to mention the crossbreeding that can bring together complementary abilities that, when combined, generate new things. As we have seen, the main selection pressure that favors language is, first and foremost, technical development, whether corporal or external, insofar as it requires an increasingly precise description of the objects that distinguish the end, the result to be achieved, the means to be implemented; it is a question of identifying what we are talking about. To speak of ends, of objectives, presupposes that each recognizes the intentionality of the other. Socialization and technical development, which have been constantly improving and becoming more complex since the beginning of verticalization, lead to ever-increasing pressure for language modalities to emerge.

Bergson believed that language arose to meet communication needs in the face of danger, even though all animals find ways to communicate in this regard without articulated language, including higher primates, such as the already socialized chimpanzee and bonobo. The latter are able to communicate with each other without language. It was Pradines who first identified the needs of using and maintaining the technique as one of the most important conditions for selecting a language.

At the same time, André Leroi-Gourhan (1964), using a paleontological approach, concludes: "Man makes concrete tools and symbols, both of which are part of the same process or rather use the same fundamental equipment in the brain. [...] Language is possible from the moment prehistory delivers tools, since tools and language are neurologically linked." Taken literally, this would imply that language could have appeared with the first tools, that is with the *Australopithecus*. However, this is difficult to support, because in the *Australopithecus*, the deficiencies concerned the anatomy of articulated sound production, including the shape of the palate (Jeusel and Mafart 2003, 2006), breath control and numerous neural connections. Leroi-Gourhan was first interested in *Homo sapiens* and *Homo neanderthalensis*, in a context where hominization had already begun to change in nature, and he probably underestimated *Homo neanderthalensis,* so close to *Homo sapiens.*

Verticalization changed the angle of the pharynx and mouth axis from less than 30° in the chimpanzee to 90° in *Homo habilis*, providing important structure for sound modulation, while the position of the glottis and hyoid bone became critical for sound production (see the bellow of a deer). In order for sounds to be produced and modulated correctly, fine control of breath production is required, which in turn depends, among other things, on the central nervous system. *Australopithecus* had to vocalize better than a chimpanzee and even more so than *Homo habilis*. The analysis by Jean Granat and Évelyne Peyre (2004) shows that, in fact, in the lineage of Homininae, everything is already positioned for possible speech, anatomically speaking, thanks to the combination of verticality, which changes the angle of the pharynx and mouth axis, and the low separation between respiration and food absorption (descent of the glottis). "Representatives of the genus *Homo* had a hyoid bone and a larynx, of which the positions and dimensions are close to those of the present Men. Their vocal tract gave them the potential to speak." The position of the hyoid bone and larynx is not a consequence of the need to speak. On the other hand, this need allowed a selection in the very long term of this putative phonatory tool.

Lieberman *et al.* (1984) identified the appearance of language from "the descent of the larynx", commonly referred to as the "descent of the glottis"[2]; this assertion is questionable. For there to be an evolution that leads to speech, the phonatory apparatus must have existed beforehand, for other reasons. Producing a language is more complex than a simple "descent of the glottis" and requires a whole evolutionary process in the brain, jaw, mouth,

2 This "glottis descent" theory is clearly Lamarckian and not very credible.

etc. Here again, we have reasoned backwards in a Lamarckian way, taking the *ex post* result for an *ex ante* objective. This appears now clearly (Nishimura 2018).

The development of bodily technicization and external technique, equipped, up to *Homo habilis*, could be realized without the need for language as we know it. If, the position of the pharynx having already been lowered, it is immediately affirmed that all the means exist to speak, it is because we consider that all the other functions and conditions of speech have been put in place. How? If the nervous system allowing the control of the articulations (jaw, lips, tongue, palate) is not in place, it is not the ability to produce sounds that will suffice for an articulated language and vice versa. Nevertheless, this makes vocalization possible.

There are three broad categories of possibilities. The two systems, neuro-articular and pharyngeal descent, take place independently for different reasons, in one order or the other, or they take place approximately together, over time, by retroactive loops. If they take place together, it will be necessary to understand how the whole system is selected. In contrast, the explanation by Jean Granat and Évelyne Peyre is simple: everything is ready to produce sounds and this is due to verticalization. The question becomes only the order of the selected functions that will contribute to language and that can be separated into three: the mastery of breathing; the profound modification of the musculo-maxillary system, modifying the shape of palate; and the neurological control of the production of articulated sounds. This requires a selection pressure that promotes a more fluid and variable pre-speech.

At the beginnings of the technique, it is conceivable that rudimentary means were sufficient: look and copy. The bonobo Kanzi copies the activity of obtaining a flame and starting a fire. To speak, it specifically requires a control of the breath, voluntary or semi-voluntary, and this ability is physiologically different from that of the articulations of language. We can conclude that for over a million years fire was used before smooth and fluent speech appeared. Maintaining fire requires controlled breathing. It is enough to look at young children blowing on a fire, or to remember our early childhood, to see that in front of a fire that is going out, trying to blow and adding wood is almost spontaneous. Did the men of that time blow on the fire with hollow sticks, or with hollow wood?

Well directed, the breath is more efficient; the use, and therefore the selection, of a bodily technique before the equipped technique, when the beginnings of body capacities were present and variable, seems more plausible. Blowing on a fire in a directed manner is an advanced body technique, as is verticality. While fire was used a million years before speech appeared, we can imagine that the ability to blow on a fire was a selected advantage: the low position of the glottis was selected for this control, all the more so as there was variability at this level.

The production of modular sounds, allowed by breath control, may have been an exaptation, as if described for bipedalism. The aptitude for controlled breathing was a condition for the selection of the positioning of the glottis and hyoid bone, which, according to this hypothesis, results from the use and maintenance of fire. A chimpanzee is able to blow, but it is the control of breath retention and regularity that is lacking. From the moment fire exists, its maintenance, which leads to improvements to voluntary breath control, can have many other effects, such as access to rhythm and modified states of consciousness (Orr 1977; Grof 2010). Such an experience can become the basis for new behavior.

He who can blow on a fire, in a more stable and continuous way, will support his group. One can admit that this creates the conditions for the selection of the glottis control (which separates the respiratory and alimentary canal) and then makes it possible to use the joints that movable lips, tongue and other muscles of the mouth and throat make possible. The closure of the glottis during swallowing, caused by the upward movement of the larynx, is not voluntarily controlled. It is controlled during breath retention, which every human is capable of achieving before the age of four, whereas every chimpanzee has the greatest difficulty in achieving this. The control of the glottis allows precise sounds to be emitted.

Breath control has another advantage, in that it is an amplifier of verticalization. It is not for nothing that spiritual masters in the East insist so much on both verticality and breath control. Anyone will find that working with the breath to achieve meditation is easier lying down than standing or sitting. Standing stiff, in a martial posture, is not conducive to a stable and ample breath. Mastering both, verticality and controlled breath, remains a basic secret for martial arts. We can conclude that the flexible and relaxed verticality is one of the last links for the human reach. *Homo erectus* had almost everything, including the oolitic organs of the inner ear reacting to linear accelerations and the semi-circular channels of the labyrinth to angular accelerations (Skoyles 2006). The path to the modern human was probably more about neurological integration.

Domesticated fire was a protective gathering place, especially in cold weather or at night; the need for exchange must have increased, creating selection conditions that favored the emergence of an articulated language, especially for more or less nomadic groups. If the positioning of the hyoid bone was correct and adjustable, singing became possible. The modulated song first appeared, then the articulated song emitted at night around the fire, scaring away predators. This is Darwin's hypothesis, whose multiple premonitions and intuitions give an idea of his genius. Thus, a protolanguage was created long before speech could fully develop, with grammar being constructed from a language of the hands, for example, practiced long before spoken language (Bickerton 1995) and necessary for whistling languages (Meyer 2008). Any variation in the functioning of the pharynx and glottis, improving sound production, would thus have been selected by social conditions and the ability to maintain fire, despite its disadvantages for swallowing and breathing.

We see here that the argument of glottis descent in the dugong, sea lion or deer does not imply a similar selection. These are convergences which confirm that this biological process exists; in the former, it is a selection on the ability to emit powerful sounds. In *Homo*, it may have been driven by the need to control fire. More than for other biological phenomena, the appearance of language could be a succession of exaptations, because of its complexity. The advantages of language are so numerous and varied, from the moment when life in a group is complex and requires many techniques, that there could be several paths leading to its appearance. We know that naming our fears and anxieties allows us to reduce our phobias; did the fireguard at night, with its ghostly lights and shadows, awaken anxieties and wonders that our ancestors needed to share?

Being able to choke on food does not appear to be a biological advantage. On the other hand, the social and technical advantage of breath control is such that the direct biological disadvantages become negligible. The social and sociotechnical relational system, which protects against biological fragility, does more than compensate for the biological deficiency: it allows it, and in some aspects, may even favor it. Breath control, which leads to experiences of altered consciousness, could be the seed of beliefs that can be qualified as shamanistic. They were dominant long before agriculture, even before the break-up of life in the 6th Century BCE and have not yet disappeared.

Those who speak with sign language mobilize the left and right brain. Some researchers even believe that the right brain is involved in metaphors, figurative aspects of language and intonations. All these hypotheses are widely discussed by researchers. Do we not see that it is the technique, rather bodily technique, in its plurality which leads to a different selective process, whether it is the control of the breath or the cerebral control of the different muscles of the phonatory system? As soon as the ability to emit sounds or to make complex signs with the hands, for a representative language, began to appear, it was selected by the technical advantage it brought to the group, all the more so as the ability to produce a wide range of sounds was practiced.

Talking is a technique; it can be learned and it can be transmitted. The possibility of an almost infinite number of languages with multiple grammar systems shows the technicality of language. Some are more gifted than others, as is the case with all techniques. Speaking a specific language can be learned as well; a person can speak several languages. One of the rare bards with celebrated talent, whose name has gone down in history, long before Shakespeare, is Homer. The magic that we feel when faced with the mastery of the verb is similar to the magic that we feel when faced with the mastery revealed by an acrobat, a dancer, a musician, a sportsman. Any exceptional technical mastery is fascinating. It would be difficult to understand why a population that has mastered many bodily techniques and many tool techniques, and that is constantly improving them under a constant and multiform selection pressure, would not improve the body technique of the language from the moment it begins with poorly articulated sounds and/or hesitant tones. Because language, by appearing after everything concerning phonation, breathing, and even part of the technical mastery of the labial, lingual, velar or glottal articulations, can be subjected to a selection pressure: the need for transmission-learning of the techniques used, the need for more precise exchanges on intentions and reasons, dreams and desires, hunting plans, worries, as well as the existence of a night life near a fire. Everything existed to select a more effective language. Jean-Louis Dessalles (2000) has shown that the final selection could be the need to create strategic alliances in increasingly numerous groups.

In a sociotechnical context such as that from which *Homo sapiens* emerged, general bodily technical mastery, including vocalization and sound articulation, encompasses the technical abilities specific to language. Language was the last vector favoring the general transmission of techniques and their proliferation from this period onwards, which has been described as a "human revolution", or a "cognitive revolution", that is during the period prior to 50,000 BP. Chris Stringer (2012) shows that the encephalization coefficient began to increase after the meat diet and tool making, that is after hunting. He shows that the African Middle Stone Age, which ranged from at least 200,000 BP to 50,000 BP, opened up a new humanity that would show in its artefacts the capacities that we imagine to be those of humanity, taking into account the fact that the populations of the time were not sedentary and few in number. We can deduce that, as in the case of walking, the spoken word must have emerged hesitantly and approximately. It became accomplished during this period. Let us recall that it took more than two million years for the vertical posture to be anatomically identifiable, and as much for a balanced, supple and ample vertical walk. That it took tens, or hundreds, of millennia to finalize a doubly articulated language accessible to learning, even with the neurophysiological and anatomical bases now present, does not seem excessive. Moreover, for it to be marked in the paleontological archives, it still needs additional time for expansion and production, as well as adequate conservation conditions.

In summary: tools and manual skills (the baby carrier, cooking, child protection, collective actions, intra and intersex technical exchanges, etc.), which *homo erectus* knew how to transmit from generation to generation, created conditions for selecting brain development and reducing the length of the child birthing process, through better child protection. In interaction with this development, an enriched social life and a more or less efficient protolanguage, with the hands or various sounds, created a selection pressure on a more efficient language. The domestication of fire provoked a selection on breath control, made operational, for speech in the context of subsequent brain development. This led to a longer larynx; singing became possible, accompanying the development of the sacred, the metaphorical, the signifier, which strengthened the group. A group that sang and danced around the fire was more attractive; it is known that in hominids, it is the females who leave their group to join another one. The party around the fire may have been an element of selection towards humankind. Even today, singing, talking and dancing, warmth and night lighting are the dominant ingredients of attraction, including sexual attraction. Singing and dancing, moreover, seem to be so deeply rooted in the depths of the human being that it is quite possible that for thousands of centuries it has been an activity of *Homo* (*erectus* and others), at night, around the fire.

It is then, archeological records show us, that the brain increased by 30%, exceeding 1,500 cm^3; *Homo sapiens* and *Homo neanderthalensis* entered the scene; at the cerebral level, especially in the prefrontal and frontal cortex, everything seems to exist anatomically. Pretending that language came next seems unrealistic. If it took a larger brain to act, speak, sing, think, conceive, reflect better, it was action, speech and singing which were the selective conditions for this improvement. It was more difficult to articulate well while singing, so the selection pressure may have been stronger. Let us imagine night scenes around the fire, songs that chased away predators as well as fears, collective and danced songs that already spoke of the mysteries of existence, linking night to daily experiences, metaphors of daytime life, finding in the inner images what reveals the meaning of those of the days[3]. The collective imagination could be staged, competitions could appear: sacred collective games and shamanic rituals. The groups with the best singers, the best blowers, the best dancers, had a selective advantage, then those with the best daytime storytellers, because the group became more united. The capacity for language allowed for continuous learning and innovation: language was being developed over many generations.

This was done over a long period of time. Between the full spread of *Homo erectus,* the expansion of *Homo heidelbergensis* and the appearance of *Homo sapiens*, there was at least 800,000 years. However, a simple mutation, which appeared about 500,000 years ago, that is 200,000 years before *Homo sapiens* was recorded in paleontological archives, led to a considerable multiplication of brain neurons (Florio *et al.* 2016, 2017). In areas of good skeletal conservation, *Homo sapiens* could be found older than supposed.

---

3 The foundation of language is metaphorical and we forget this, so far away are the origins of today's words. We need only look at how new words and concepts appear, even in science (Lakoff and Johnson 2003). "Without analogy there is no concept, and without concept there is no thought" (Hofstadter and Sander 2013). Epistemologists who rail against metaphor do so according to metaphorical expressions of which they are unaware: "it just jumps out at you!" What about "clear and distinct thinking"? About "chain of reasoning"? Bachelard (2001) writes: "A science that accepts images is [...] a victim of metaphors. The scientific mind must therefore constantly fight against images...."

About 300,000 years ago, the brain volume had reached the level we know, even a little beyond. Speech must have been made by trial and error and not at the level of modern humans. Indeed, if ancestors called "anatomically modern" appear around this time, nothing can distinguish their tools from other *Homo*; they lived like them. The big brain selected, for evolutionary reasons that are both affective-emotional-relational and technical (tool, body, mind), was not much more efficient, technically speaking, in the immediate future. It had to play a role in the quality of night life. On the other hand, it seems coherent to think that the variability of its operationality should be great. This leads to an increased power of selection through language, a selection process that has become primarily psychosocial.

It is during a period from 70,000 to 50,000 BP that a considerable transformation is detected in the paleontology archives, which means that the transformation took place long before 70,000 BP, during the so-called Middle Stone Age. Molecular biology suggests that the final corresponding population – from which today's humanity originates – was reduced to the order of tens of thousands of individuals. Curtis Marean (2015) shows, however, that population size plays an important role in maintaining a complex technical system, as the division of labor increases even more. It is quite conceivable that a population of 10,000 or 20,000 individuals was sufficient to maintain all the techniques of the time. This means that the last wave to come out of Africa came from humans who were capable of all adaptations and were very cooperative in the face of the planet's multiple environments. It overwhelmed the world while mixing with a few more or less archaic *Homo* already present there, far fewer in number. This scenario seems coherent.

Chimpanzees, which amounted to more than a million at the beginning of the 20th Century, were estimated to amount to a little less than 200,000 by 2019. They are certainly threatened; this means that in the last stage of the hominization process, the threat to *Homo sapiens* was even greater. Because for any collective, innovation has a cost: possible extinction. This is why successful developments are bush-like. On an evolutionary path, one success makes up for ten failures. When metastable conditions do not favor a branch-like evolutionary structure (i.e. the opening up of possibilities) the probability of failure increases. Conversely, a stabilized, non-evolutionary species is a species without history. In some cases it lasts for a very long time (e.g. coelacanth or gingko); we must be lucky enough to be in a stable, ultra-protected "niche".

According to the reasoning I have followed from the beginning, what happened before that period seems conceivable. The selection of a large brain may have been the effect of esthetic, emotional, relational criteria. The baby with a big head, overdue and crying, may have led to centration, an additional group isolation; that is to say, in terms internal to human groups, an increased cooperation. In any case, the group had the means to live and develop, because the mastery of fire and the classical techniques of *Homo erectus* allowed for survival in an environment with diverse resources and an increased capacity to use its environment, however variable it might be.

The additional capacities of these humans with more elaborate brains, in a few thousand generations, led to incremental changes, both technical and verbal, and therefore conceptual, which, at the end of this long journey, would lead to the creation of a system: easy transmission of techniques, design of new ones, sociotechnical and cultural integration. These new forms of knowledge and know-how are, in the evolutionary sense, contingent, unpredictable exaptations. The large human brain, which gave way to "the modern anatomically human species", appeared well before the end of the *Homo erectus* period and before this human revolution (see above) which probably began more than 100,000 years after the acquisition of this brain – whose differentiation could only be built up in interaction/interrelation over many generations. At 70,000 BP, a familiar humanity appeared. It came to leave marks preserved in the sumptuous caves of the Aurignacian and Solutrean. It was a humanity that even tried to expose its inner life on the walls of the caves.

PART 2

Technique and Human Ontology

13

Technique as the Foundation of the Human Being

13.1. A look back at stone-knapping: the contribution of neuroscience

Our proposals could be considered very speculative. Experimentation would be needed and it turns out that with regard to the production of knapped stones, two American anthropologists, Kathy Schick and Nicolas Toth, proposed, as early as 1990, an observation of the cerebral functioning when knapping stones. This work has been pursued by several researchers, and their conclusions are in line with the logic of the arguments we have presented from the beginning (Shea 2016). Learning stone knapping according to the Oldowayan method already required tens of hours of concentration, while the late Acheulean method, capable of producing cut axes, required up to 300 hours of learning for a modern *Homo sapiens.* The so-called Levallois method, dating from the Mousterian to the Middle Paleolithic, was more conceptual and cannot be transmitted without explanation.

The correlation established with brain activity helps us to understand the nature of the difficulties. The most difficult and demanding brain activity was stone knapping. The force of the strike needed to be controlled according to the stone and tool being cut; the precision was approximately a millimeter. As with chisel carving, fragments were removed with each cut and could not be put back into place: one mistake and the whole work was ruined.

Sharing attention was the genius intuition of Pradines; it was about keeping the goal in mind and adjusting the strength and accuracy of the blows. This control required a long learning curve. It has been clarified with the help of brain imaging techniques. Oldowayan pruning mobilized the premotor ventral cortex, the anterior and superior intraparietal sulci and the supramarginal gyrus of the inferior parietal lobe, the latter being known to be involved in the awareness of the body's position in its spatial environment[1]. The requirements of the work completed were advanced and included additional mobilization of the superior premotor cortex, involved in manual tasks and the left lower frontal gyrus involved in the control of retrieval of information from long-term memory.

It required both planning thinking and precise motor control. The Levallois (Mousterian) technique was surprisingly complex; Éric Boëda was able to study it and present the "Levallois concept". One can only be in awe of such ingenuity. Researchers have shown that this activity led to a detectable modification of neural networks. During any learning activity, a neuronal reconfiguration takes place in the process. This defines new behaviors that will be subject to selection, favoring the most efficient variations. Tool use was indeed a driving force of brain development. The cutting of a flint required a continuous and thankless effort, rewarded by ... the pleasure of success; this was amplified by the need to be guided and supported. It was a collective activity. It required sustained control and concentration. Every stone-knapper, even today, knows that it is necessary to understand the characteristics of the stone to be cut. This required a long apprenticeship that needed to be transmitted (Firth *et al.* 2009; Hecht *et al.* 2015; Morgan *et al.* 2015; Stout *et al.* 2015). The brain had thus acquired its current plasticity before the Mousterian, allowing for the most advanced stone-knapping methods to be learned (Gärdenfors and Högberg 2017).

After verticality, the selection pressure promoting a more efficient gait led to the widening of the pelvis, indirectly allowing for an easier delivery of children. A selection pressure, through technique, leading to the development of the brain thus became effective. Neuronal plasticity, already well explained by Jean-Pierre Changeux (1985), means that the same area can be mobilized for various activities. This is the case of Broca's area (lower left gyrus), often described as the language area, and also mobilized

1 The superior intraparietal sulcus supports visual short-term memory for multifeature objects and the anterior intraparietal sulcus supports goal representation.

for music, mathematics, as well as in actions requiring manual dexterity. We can therefore suppose that what has been selected by the tooling technique has also been mobilized for language, whether gestural or oral. What must be grasped is that it also works in the other direction, further proof that language is a technique. With the invention of writing, another neural mobilization is realized (Dehaene 2009). The technical association between body and mind (speech) is externalized by a device that combines bodily and tool technique (writing).

It is always the technique that creates new selection pressure which guides evolution. If an evolution seems slow, like the one that took place from the first vertical primates to *Homo habilis*, it is because, contrary to what has been estimated by some, the technique of stone-knapping, associated with verticalization and hunting, required multiple capacities in complex interaction. It was a transformation that required many neural rearrangements over generations.

Even while at the beginning, this "knapping" of stone was implemented according to a Darwinian process of trial-and-error with copying, the result gradually led to more and more systematic approaches. Each success was stabilized by copying between individuals. This meant that the collective learning of verticalization selected copying abilities that higher primates did not have, without visible brain growth. The use of stones from the pebble culture, random stone breaking, the Lomekwien and then the Oldowayen selected more advanced abilities. At this last stage, the ability to understand what was being copied was more advanced. Stone knapping required a control of one's gestures and a choice of materials that amounted to a kind of anticipation, albeit approximate. From the time of *Homo habilis*, stones travelled to be cut in competent centers (Putt 2015).

The quality of the stones used was a strong selective advantage in the use of these stones, for hunting or retrieving food; being able to make such objects meant being able to act in a targeted, precise, adapted way, even if skills were already being shared at that time. Just as knowing how to write gave an undeniable general advantage, so, at that time, knowing how to cut stones conferred a certain superiority. In the long run, this produced, by selection, groups that knew how to learn by copying and how to improve the method at the same time. From generation to generation, the pressure of selection led to anticipation, reflection, manual precision, the adequacy of the reasons for the methods and purposes. Each improvement was a new selection pressure for higher cerebral capacities in terms of control of gestures, control of precision, matching it to the perceived objective to be reached and to the choice of material. The choice of the latter led to searching for the "good stones" and to transporting them to the knapping location. Collective coordination, task sharing and anticipation of what was expected was required.

We understand that it took more than two million years to go from the random break to the Acheulean method: a level such that to learn it today would take several hundred hours.

13.2. Explaining humans through technique: a conceptual error

Technique modifies conditions of existence, conditions which, in the long run, are as many selections of the populations that use it. Any shift in a population's conditions of existence creates a selection which, in the long run, leads to an evolution of the populations from which it originates. This is not specific to humans. What evolves are more or less large groups that can bring a species together, as long as inter-fertility between groups occurs and is maintained.

Human nature is a consequence of the main "driving force" of its evolution: the technique: physical and equipped, then mental. On the one hand, it changes the conditions of existence and, on the other, it is produced by the group itself. We are in an unstable/metastable system, because the technique mastered at a given moment, from the beginning, has a selective effect, in the long run, on the biology of those who master it. It creates a selection pressure which directs towards a population with capacities closer to a better use of the technique for the development of groups, which requires a more precise coupling between its conditions of use and the consequences of this use. This adaptation becomes, in fact, a selection pressure for the appearance of a more efficient technique or technician. If the natural environment is changing both in space and time, the technique will be selected according to its ability to allow the group to adapt and survive in extreme conditions. It is the characteristic of the technique to offer greater means of immediate adaptation, thanks to the technicized environment[2],

2 The word used in French is "*milieu*", this part of the objective environment which is adapted to and recognized by the species.

which acts as a mediator between the environment and humans (or other technical living beings).

Thus, it seems as futile to seek the birth of the human being outside Africa as it is to seek it in a specific place in Africa, except perhaps in the very last stage. This continent, at the origin of both panins and hominins, offered all the conditions for differentiated evolution, without however causing isolation and too brutal a break (as, for example, the glaciations in Eurasia). The various exits from Africa have almost all failed over time, sometimes over a long period of time, except that of *Homo sapiens,* from 50,000 BP, a species which could adapt, technically, to almost any climate on the planet. Questions about sustainability will now follow.

Predicting what a new technique will select is a challenge. This is the contingency principle of evolution; technique does not change that. Nothing prevents us from imagining the possibilities it reveals and confronting them with reality. In the bush of the expression of possibilities, there is at least the case of naturalization or renaturalization, a form of biological respecialization. The most cited case is that of *Paranthropus boisei.* Adapting to a vegetable diet, and coming from the lineage which came from verticalization, it was a contemporary of *Homo ergaster*, *Homo habilis* or *Homo rudolfensis*. It had not succeeded in adapting to a climate that deprived it of its specialized food possibilities. The "un-tooling" or "de-technicization", always possible, in the lineages that led to Man, and that led to environmental specialization, were dead ends while the environment changed quickly. In relation to the dead end, *Homo habilis* was to *Homo erectus* what *Homo erectus* was to *Homo sapiens*, an ending, but with an earlier branch coming back to life.

To speak of an explanation of the evolution that leads to humans through technique is therefore excessive. Technique does not explain evolution, it opens up possibilities to a species that has them, whether these possibilities are already underlying or whether they appear through mutations of all kinds. This technique is at least two-fold: corporeal and equipped. From a stage of phylogenesis, around *Homo erectus* and *Homo heidelbergensis*, it tripled, because the mental technique was added. It was built by the need to identify the reasons, motives and ends; this need was accentuated by the necessity to transmit and maintain collective knowledge. This resulted in a harsh selection. Later on, starting from the Mesolithic, new families of techniques were put in place.

The question evolves: why is it, in the hominin lineages, and up to today's human, the technical advantage promoting biological evolution was also favored by it? The first answer is that, in an immense, varied and changing environment, technique is favored because it offers protection, the creation of a reconstructed and stabilized environment, and a more efficient use of the non-humanized environment: every living being becomes prey. Thus, the multiple technical innovations improving these three parameters are both new conditions of biological evolution which are selected by the use of tools, and conditions of demographic growth in the environment. A positive feedback loop has been established between technique and biology. A comparison with other animal societies would be interesting.

Protection by the technique modifies the physiological needs whose variability is selected in relation to what the technique provides, as well as the capacities of use of the technique which are also variable. We are witnessing an apparently paradoxical evolution. The technique de-specializes the species which develops it, as we have seen with the various *Australopithecus*; the one that was in the lineage that led to humans being more slender and more undifferentiated in terms of diet. With the help of tools, it could hunt and eat meat. Although more graceful, it was nevertheless a more efficient predator.

It can be said that, through the bodily technique, however "minimal" it may be – that is, originally, verticality, the freeing of the hands and the use of simple tools – the body is put in a position to be selectable by the tool technique. The latter became the first media between man and his environment, in the same precise sense as Simondon's, in which the tool is a mediator between man and his environment. Here we can take up the concepts proposed by Augustin Berque (2015, p. 206): "Mediance is indeed the structural moment established by the bipartition, specific to the human being, between an animal body and a medial body." The "medial body" is one which is capable of bodily technique. Hominization, which began with the learned and transmitted verticalization, revealed the beginning of a mediance. The *Australopithecus* created its own protective and at the same time selective environment. This developed during the whole process that led to *Homo sapiens*.

As a result, there was both a bush-like evolutionary structure and the impression of directivity: a lineage is differentiated by a better biological adaptation to the tools, a capacity for technical innovation allowing access to a richer and more varied diet and the protection of children and mothers. The feedback system being positive, the "collective system" is in fact moving in the same direction, even if many possibilities lead to diverse species that have not triggered the positive feedback and stop there.

It is a probabilistic function that is all the more efficient as it produces variations. The cost of any probabilistic function is evaluated by the losses of unsuitable variations, compensated by a higher average efficiency and the possibility, sometimes, that a new path may appear. The reason for this is that a system is maintained in the averages, but evolves through the extremes, to paraphrase Valéry. This does not prevent stasis, in accordance with the theory of punctuated equilibria, which can give the impression that nothing changes for a certain period of time. Within the evolutionary trend that led to humans, stasis can be interpreted as a specific difficulty of biological evolution. Increasing the brain and/or its capacities depended on both the shape of the hips and the pelvic opening, the decrease in size of the jaw muscles, the protection of babies at birth, the rebalancing of the body, mutations of homeotic genes, etc. This took time, even if it was a response to constant selection pressure, because it was contingent, depending on random selectable mutations.

Developing the spoken word required even more changes. One can thus understand the long period from *Australopithecus* to *Homo habilis,* and that of the beginning from *Homo erectus* to *Homo sapiens*. It is the same analysis that makes it possible to grasp the stasis between anatomically modern *Homo sapiens* (before 300,000 BP) and the emergence of more creative human collectives (between 70,000 and 50,000 BP). Transformations require either sequences of innovation, or mixes, or both. This was the same rule in intellectual conception or artistic creation during the historical times of *Homo sapiens.* A new concept or a new artistic approach could only appear when conditions made it possible and when there was both variability in the capacities of individuals and a selection pressure favoring creation.

Michel Foucault (1994) demonstrates this in terms of the mental concept; to conceive a new idea, we need intellectual conditions that make it possible to conceive it, that this thought can be accepted, taken up, developed. Not everything is possible at all times. A new concept, or a new practice, or a new brain process opens the way by blowing up a kind of "limiting factor". The answer here is the equivalent of a theory of encounter (Morizot 2016).

The new is an event that is the result of a contingent encounter between a new possibility and a function (or structure) that can seize it.

We can simplify the model. The techniques, both physical and instrumental, in interaction, or even entanglement, were the main driving forces behind co-evolution; the variability of the environment, external to the specific environment created by the technique, was the selection system that favored the technicity; the capacity for predation (hunting, gathering, capacity to process the products that were food sources) was favored. The emergence of human beings, in their ecological position, was not unfinished business, on the contrary. It was the full completion of populations made for the optimization of hunting and gathering. Since 50,000 BP, no species can resist human predation any longer. Global expansion was a consequence. It even seems astonishing that it did not go faster. This could be blamed on infectious diseases. It remains to be analyzed what this feeling of incompleteness of the human being, presented in reference to neoteny, means (Dany-Robert Dufour 2012).

This model is probabilistic: we cannot predict the nature of the variability or the nature of the variability of the transformations; we only know that when a selection pressure is caused by the technique and its use, it would select, in the long run, mutations that opened up the field of possibilities in a certain way. There were a more or less limited number of possibilities, and a selection (or choice) eventually responded to them. Stasis can be thought of as the succession of punctual improving processes and the bush-like evolution as the consequence of a new opening allowing multiple paths. The bush-like evolution is the indication of a disruptive innovation.

13.3. Mental exaptation as a norm of human development

Sloterdijk notes that transposal capacity is a human capacity that distinguishes humans from other species. It is understandable that the practice of technique, and its incessant use, may have selected such a capacity. We have seen this with the neuronal development of Broca's area, which had much wider consequences. It became very widespread. Transposition, or exaptation in evolutionary biology, is not unique to humans. What is specific to it is its generalization, which takes place at the cognitive and behavioral level. Being able to transpose means being able to discover new uses for different situations based on common patterns. The use of techniques leads to the selection of this aptitude, if it can appear, which was the case. We can, *ex post,* understand it: exaptation having been "used" by living beings many times, we do not see why it would not have continued. The novelty comes from the selection pressure exerted by technical use. It favors a cerebral development which will lead to the development of the capacity of generalized transposition, also based on a technical bodily capacity.

The cerebral synchronized reentrance described by Edelman (1990, 2004), by generalizing in successive stages, eventually led to reflexive consciousness[3]. And there is no reflexive consciousness that is not exaptational; it is in transposition, metaphorical or analogical. Generally speaking, exaptation coincides quite well with changes that seem important after the fact and that seem to escape logic. Indeed, the process resembles what Simondon calls transduction, that is to say, a "reasoning" that follows forms, contours, links, relationships, connectedness and kinship.

From the *Australopithecus*, biological and behavioral exaptation was selected. It also became mental through the exercise of bodily techniques and tools favorable for survival. It was a mark of the decisive influence of the use of technique. The hand, for example, thanks to technique, was able to experience an ever-increasing range of uses. The mouth, an organ of grasping and aggression, came to be used for expressing feelings, speaking, kissing, etc. The slowness of the process was due to the slow pace of selectable biological transformations. Once exaptability became a mental capacity, another form of adaptation became possible. To the extent that the body supports it, any form of exaptability can be expressed. Its modalities of expression are sociotechnical rather than biological, because the body and the neuron have acquired a flexibility that goes far beyond the strictly biological possibilities of innovation, which have been externalized by technique. When exaptation, or transposition, became the norm, it partly freed itself from genetics. Neuronal plasticity is such that the brain, in each generation, is the result of a kind of slightly modified reprogramming. The evolutionary rhythm can be accelerated; each generation has the potential to experience a technical break. This does not eliminate biological selection, but displaces it even less predictably because it is not the strongest but the most flexible that could be selected.

3 Gerald Edelman's general thesis can be considered as neuronal Darwinism; he was awarded the Nobel Prize for his studies in immunology, where, again, he applied Darwinian concepts.

Intentionality is selected by the use of the tools and projected onto the tool; it becomes exapt; it becomes permanent exaptation. As soon as the technique becomes quite complex (before *Homo erectus*?), it becomes a condition for the selection of the capacity of transposition which is *de facto* exaptation; it becomes the rule and not the exception, because any capacity to develop a more efficient technique is favored and can be realized from mental patterns directing the body according to the principle of a corporal technique. The evolution towards humans was driven by the technique, even if the mind also took over the control levers of the selection processes.

Paths can be crooked. The reduction in size of the chewing and digestive system, which took place during the time of *Homo habilis*, was promoted by the use of the technique and not caused by it. It corresponded to an improvement in the vertical position, and a decrease in biological energy expenditure, which indirectly paved the way for potential cerebral growth. The latter is no longer subject to negative selective pressure (maxillary muscles which were initially fixed on the top of the skull).

The human body has thus been modified by technical use, over several million years, as well as its behavior, more adapted to technical use. Cognitive capacities, whose variations have been selected according to their correspondence to the requirements of technical mastery, are themselves constructed according to technical logic. Today, it still seems impossible to predict and explain the psychic emergence; we can only note it. Similarly, we note that primates have a brain that is more developed than other animal species (except cetaceans); this is due to both the importance of social relations in these species and to the complexity of the environments in which they live. The specificity of the psyche is that it cannot be analyzed according to the technical approach, whereas the cognitive capacities are. For the cognitive capacities are linked to the fact of being able to learn and teach; it is technical internalization. However, this distinction can be discussed and disputed.

The development of tool techniques reinforced the trend towards a technicization of the body and mind. This can be attributed to the necessary cooperation in technical activities that provided access to food. It is excessive and unrealistic to speak of technique without reference to the social aspect, although to speak of the social aspect without reference to technique is worse. Social cooperation not only promoted technical development, but in its very definition it is an irreducible component of it.

Without a relationship, there is no technique, which does not mean that the relationship is essentially technical or that technique cannot function without a relationship. It is ethically preferable that technique implies a relationship. But this is another matter of great importance that took place after hominization.

Technical cooperation, in and through the relationship, means specializing at the individual level. This specialization means that each individual has a potential that is broader than what he or she is going to develop; the act, in fact, removes possibilities. Everyone finds a specific place. It is a system that is set up and can be described in different complementary ways, particularly from a technical or social point of view, and also, *in fine,* from a mental or psychological point of view. We can speak of a sociotechnical system very early in this evolution. That said, if all this text is placed under the technical angle, it is because this aspect seems to me to have barely been studied historically; not all humans or everything human is technical. But the whole of humanity and everything human is based on, is supported by, the technique. The human dream of being able to acquire all the forces and powers that come from technique, without technique, is a dream of omnipotence. It is the dream of the superhero who almost needs no one and who frees himself from his own destiny.

The generalization of exaptation in human evolution is translated in *Homo sapiens* by a unique capacity: mental exaptation, that is the use of analogy and metaphor. A characteristic that appeared to be biological manifests itself in and through psychic processes, that is through neural processes. In fact, all figures of rhetoric are particular cases of a more general technical function which is the essence of thought. We have already shown its importance in the formation of thought and language (Lakoff and Johnson 1986; Hofstadter and Sander 2013). It is an exaptation by reflexivity, which has become mental/neurological. Metaphor is an extension of analogy, a technician's analogy, into the psychic field, which is that of signification and the construction of meaning (Dubois 2017).

13.4. The relationship between bodily technique and tool technique

The theory defended here is that what was the catalyst at the beginning of this long evolutionary process of more than seven million years was the appearance of the first bodily technique for permanent use. Here there was no innovation of a new tool technique. It was the acquisition of a particular learning process, the verticalization of the body and bipedal walking. The gaze became oriented in a perpendicular and ventral direction in relation to the longitudinal axis of the body. It was within the framework of this particular technique, and in co-evolution, that the learning of tool technique was developed: evolution of the hands and feet, better visibility, precision of gestures. We know that the first bipedal primates walked hesitantly, unsteadily and unstably, and that the body was not well adapted to it. This posture still has a specific cost, today, after millions of years of various evolutions. It was created in collective conditions that stimulated an incessant dynamism in bipedal beings. The whole body evolved towards more endurance, a permanent tonus, higher cerebral energy needs, increased child protection. It is not for nothing that humans need "physical activities"; it is through them that their mental capacities have developed.

There can be no elaborate tool technique without a bodily technique, unless the adaptation to the tool is biological, such as the snail's body is adapted to its shell, the spider's body to its web, etc. If the new tool is developed for a particular use that needs to be learned, it is the ability to learn that will be selected. The development, manufacture and use of tools, as far as they are learned, lead to a selective environment of the bodily techniques to be learned and the relevant brain development. Unlearning or naturalization stops the process. In evolutionary steps towards humans, there have been many naturalizing impasses that have not been able to withstand the various climate changes and adaptation of other adaptive species through the use of tools: for example *Paranthropus boisei*, and, perhaps, *Homo neanderthalensis*.

Talking and singing are bodily techniques without any material tools. We can speak by whispering, that is to say without using our vocal chords; we can sing without words, emitting only sounds, corresponding to vowels, and even just one, that is to say without using the joints coming from the lips, the tongue, the teeth and the palate. We can blow – and breathe in – in a wide or jerky manner without using either the vocal chords or the articular system of speech. You can sing while breathing out; while breathing in, the vocal chords seem to work just as well, but the amplification effects are reduced. We speak in the exhalation, but it is possible to whisper in the inhalation as the vocal cords are not mobilized. It is also possible to whistle using very elaborate techniques that do not use the vocal chords (Meyer 2009).

The use of the breath, without the mastery of which it is impossible to speak, sing or whistle, is extremely technical. Whether it is a question of rhythm or amplitude, of insistence on exhalation or inhalation, of costal or abdominal, dorsal or ventral, nasal or oral breathing, it appears that in spite of the maintenance of self-regulated breathing, according to a basic unconscious functioning, it is possible to take control of it consciously almost at any time. It is only when the breath is mobilized by intense physical effort that it becomes difficult to control its amplitude and rhythm, although the margin for learning, even under these conditions, is important. In the technicization of the body, breath has been a late, albeit crucial, stage. In the distant future we can imagine other possible technicizations: blood circulation, digestion, differentiated mobilization of the senses, etc. But it is always in a context of use that an apprenticeship is built, and then the possible corresponding selection.

The techniques of running, jumping, swimming, diving, are purely bodily techniques. While running, like walking, is self-regulating, it is undeniable that technical mastery is acquired through learning. The succession of world records shows that the body's potential is not yet fully realized. We can see that fighting techniques can be practiced with bare hands and that they mobilize all the body's capacities. It is interesting that many combat techniques, based on the body and a minimum of weapons, have recently been developed. Martial arts, wrestling, boxing, combat sports, the innovations continue today and the current evolutionary trend is two-fold: with specific tools and without tools. This means that not all of the body's possibilities have yet been explored, more than 50,000 years after the potential was expanded.

While a number of bodily techniques require little or no tools, it can be seen that many of them do. The oldest – bipedal walking – indicates the need for specific tools: shoes. We can walk and run longer and faster with shoes than barefoot, and although prolonged barefoot running training certainly improves the body's capacities, it does not change the fact that the "walking" tool allows for greater efficiency and protects the body, as if the evolution of the foot had not been fully optimized. Shoes can be considered as prostheses that can improve the efficiency of activities as basic as walking and running. Prosthetic tools, replacing, supplementing or improving a basic practice are not new. The same goes for climbing with bare hands, but with specific footwear. This tool is here an artefact that improves body technique. It appeared, in the evolutionary process leading to man, when walking became well assured. The foot would therefore still not be ideally adapted to walking. It will never be again, because its eventual evolution is now under the selection of using shoes which themselves evolve according to multiple needs. The different walks are made possible by the use of different shoes and not by a biological evolution henceforth oriented towards the widening of the potential of adaptation to the tools. It is indeed on this specific body-tool technical relationship that prosthetic tools increasing human capacities can be designed.

The mastery of a complex set of bodily techniques leads to a new phenomenon that I describe as a theatricalization, or staging, of life and the world. The mastery of the breath, of the musculoskeletal articulations that allow speech, of the neuronal connections that allow cerebral control, and of the cerebral development that allows both reflexivity and the inner representation of what is said and heard, all of these capacities associated with the pre-existing foundations leads to the possibility of saying and showing life and the world, of recreating a situation, of telling a story, and through a narrative of reconstructing what happened. The source of this theatricalization is ancient and is rooted early on in acting, which is, at least until recently, associated with the sacred. The beginnings of what would be social life as soon as the populations were large are built from, in and with acting, the sacred, the ritual, the staging. If the theatre, in its specific organization as we know it since the 5th Century BCE, has become an organized collective technical set, what it relies on is this capacity of staging which we can assume was operational 50,000 years ago.

This new technical mastery opens up a new relational and conceptual field; what allowed it to emerge could not come from the possibilities opened up by this capacity. It is not a question of "explaining the theatre", from what it allows, but of trying to understand how it could emerge. The theatre itself, as a complete, elaborate construction, would appear long after the capacity to dramatize life had emerged. The fact that the theatricalization of life and the world leads a group to build a common representation, stable social organizations, a shared cultural whole, accentuates the cohesion of the group and creates an invisible border between groups that do not share this representation. It can be considered as one of the foundations of a fourth technical field that remains to be explored.

This dramatization is a game. It is said in most modern languages that the actors "play" their role. We find here the playful functions, preludes of the technical fields; the more these technical fields become supra-individual, the stronger the mark of the game, whose rules will have to be learned. The game becomes a means of producing a collective sense. This ability to play and to direct goes beyond everyday life, which is made up of technical details and concrete demands; one goes beyond the profane. This is undoubtedly what has led to this strong separation between the sacred and the profane. It is because the first ethnologists came from a world from which the sense of play had been altered (second half of the 19th Century and first half of the 20th Century) that they were struck by this separation between the sacred and the profane in the so-called primitive peoples. Here again, this would merit an exploration beyond the scope of this book.

The evolution of languages also created "invisible borders", and these two types of border overlap empirically. Language is a bodily technique of extraordinary complexity articulating physiology, muscle control, brain processing and social interaction. It relies on at least two important perceptual senses, hearing and sight, in addition to coenesthetic integration and overall body control over and through the mind. It establishes connections with all the other senses. As with all bodily techniques, language requires use and therefore leads to the selection of a body transformation to meet the demands of its practice. The effectiveness of language has led it to be a powerful selection tool. A recent discovery confirms this: the labiodental consonants F and V are thought to have appeared after the Neolithic period, in relation to a modification of the closure of the jaws, associated with the post-Neolithic dietary change (Blasi *et al.* 2019). The change from V to B, as in Spanish, is a kind of regression. This discovery is also a new confirmation that morphological and biological changes are quite conceivable under the selection pressure of technical use over times less than 10,000 years.

Wind instruments are extensions of singing and could meet Kapp's definition of tools as a replacement and extension of an organ. They also require techniques involving the hands, forearms and arms, and sometimes even the whole body. All musical instruments require precise body technical skills, which may involve hands, arms and even feet and legs. Playing the violin, piano, guitar or organ is certainly not using the extension of an organ, according to Kapp's theory. It is above all taking advantage of the vibratory properties of taut strings to produce sounds. It is prolonging and enriching the functions of singing according to an accompaniment mode. All those who have played a musical instrument have discovered that learning is not just intellectual. It is the whole body that has to adapt to instrument; the "domestication" of the instrument goes hand in hand with an integration, a concretization, as Simondon would say, of the relationship between the body and instrument; the whole body has to self-correlate with the characteristics of the instrument, which becomes one with the body. More often than not, the musical instrument requires a difficult bodily adaptation. Playing a musical instrument is a form of training that can be compared to a torture of the body, a self-domestication.

The same applies to all portable tools for hunting, combat, war, and production (scraper, hammer, axe, knife, saw, pliers, needle, etc.). These tools were put in place, at an early stage, by the Australopithecus; they have evolved and diversified, from *Homo habilis* to all *Homo* to *Homo sapiens*. All of these tools, including those that appeared up to the beginning of agriculture, required an elaborate bodily technique that was already possible for optimal use. It is enough to consider all the body techniques necessary for the use of a bow in an adequate way: it was almost the whole body that was concerned (Haudricourt 2010). This practice also required the ability to build the bow and arrows.

All these tools have the potential to become companions. It is a common thing to possess a tool that remains a kind of faithful and reliable collaborator, almost an accomplice, which we keep with affection, and it can differ according to culture, custom, profession (musical instrument, pilgrim's staff, "Swiss knife", toothbrush, cutlery and plate, smartphone, etc.). The fact that a tool is a reliable companion shows that the use of a tool does not only mobilize cognitive functions: it requires a relationship of trust. The user knows that he can rely on his knife, pickaxe, sword, etc. Myths have continually celebrated the sword (for example Roland and Durandal, or the epic of King Arthur selected for his ability to dislodge the sword stuck in the stone, or the famous Mo Ye sword, much more ancient, of the King of Wu). Even today, many people still do not part with their multi-purpose knife, or others with their knitting needles. Since then, the smartphone has been taking over. The affective and emotional relationship seems necessary, at least in the first contact, to know how to use a tool: "Inanimate objects, do you have a soul that attaches to our soul and forces it to love?" (Lamartine 1963).

Since high antiquity, the increased technicization of practices is most often perceived as refinement: the kitchen, the table, the clothes, even love.

Since the *Tao* of the art of love, the *Kama Sutra*, Ovid's *Ars amatoria* and the great Chinese novel *Jin ping mei*, it seems obvious that a refined culture not only does not detechnicize the human being, but on the contrary makes him more technical. The apology of a pure, supposedly natural life, is a life of austerity, detached, but never completely, from the use of tools. Everything can be learned and this learning requires concentration and method. It leads to openness to otherness, because the tool, as we have just seen, is another, neutral and friendly, human carrier. It is not a recent fact, as we have seen; it is at the foundation of the human being, so prevalent for so long that our body is marked, as well as our brain, our psyche.

Technique, as a mediating constituent of an environment, has favored brain development. Moreover, the brain itself has been selected according to this technical environment, that is in its ability to control technical activities and to be one with them. It has shown new possibilities (exaptation) that can also be interpreted from a technical point of view. Man uses his brain to a large extent as a tool. Its plasticity has become such that it is in a way the tool for doing everything; it is enough to know how to use the other organs that are objects of bodily technique in good synergy. It is a tool that now has a form of autonomy and imposes its own functional modalities. The brain as a multifunctional tool becomes the indicator of the human being, without revealing its full potential. It will take time and the resulting contingent transformation of the human environment to discover, at each new technical stage, new capacities that will be neuronal exaptations.

13.5. Variability of technical capabilities

Long before the emergence of *Homo sapiens*, a single person could not master in an optimal way all of the tool techniques and associated body techniques practiced by hominids. Even walking was poorly assured before *Homo erectus,* and it is highly probable – even if archeologists-paleo-anthropologists have not yet seen anything – that *Homo sapiens* had, during the first millennia of its appearance, amplified the range of possible techniques. Otherwise, how else would it have multiplied? There are two different aspects: the variability of levels of technical, bodily or tool competence among individuals and the broadness of the range of these tools in various uses. Different divisions of labor were established, and overall technical mastery would require an increasing number of individuals. Sufficient numbers were needed to deploy complex technical skills. It is known that around the time of *Homo habilis*, stones were transported to be cut in real workshops. It is likely that the minimum group size was a factor in the selection. Indeed, the size of the group is correlated to the variety, therefore a possibility of evolution; it is related to the number of techniques that are practiced and maintained. The more complex the techniques are, the larger the group must be to maintain them, or must it exchange know-how and products with other groups.

There must have been a significant selection pressure on the optimization of the size of a group, increasing with the evolution of the hominins. Dunbar's number defines the number of individuals with whom a person can simultaneously maintain a stable human relationship. It has been shown that this limit is inherent to the size of our neocortex. Robin Dunbar (1993a) estimated it at 148 individuals for the human species with some variability. This suggests that a group of *Homo sapiens* may be at least three times larger than a group of chimpanzees, which correlates well with the need for task sharing associated with the increasing complexity of the technical system.

The level of mastery of different body techniques is extremely varied in human populations. It is not possible to speak of a single individual who masters all body techniques to the highest level. It is simply impossible, because it is contradictory. We only have to look at athletes at the Olympics. With a little experience, it is possible to differentiate at a glance a sprinter, a marathoner, a pole vaulter, a shot putter, a swimmer, whether they are men or women. These differences are such that it is easy to imagine that, despite individual motivation, musculoskeletal imperatives specific to each sport, alone, can allow us to achieve excellence or, on the contrary, prevent us from reaching a high level. This is what coaches know.

Conversely, in a population, all individuals are mediocre, or almost incapable, in at least one body technique. The early craftsmen who made cut stone were not those who knew how to make the best use of the tools they made. This variability occurs at all levels: in a family, in a small group, a community, an ethnic group, and of course all of humanity. This variability and its distribution are recognized and affirmed by population geneticists studying human populations. We do not see why social relations would suppress these variabilities; on the contrary, they amplify or add to them.

Human genetic variability, which appears much lower than that of chimpanzees, is immense in phenotypic terms. This is easily seen in technical use, both bodily and instrumental, but it is even more important in mental techniques.

It can be assumed that much of this variability is dependent on individual and collective technical needs and that its phenotypical expression, that is the actual technical capacity of the individual, is constructed through learning. This leads to the extraordinary differentiation of populations, at different levels of scale and in different places. The selection of learning capacities, known as epigenetic, accentuates the effect of the human environment on the capacities developed by individuals.

It is understandable that exaptation becomes the dominant mode of human evolution as soon as the genus *Homo* was identified. It allows new tools to be learnt through the use of the body's technicities or already existing tools by transposition. Over the last ten millennia, a great many tools have appeared without it being possible to show that they have played a major role in biological evolution. But let's remain cautious: it was the domestication of cattle that selected populations equipped with lactase, or the domestication of wheat that selected populations capable of digesting gluten, and we have just seen that labiodental consonants appeared after a less hard diet.

There is therefore a new type of relationship between technical and human evolution in *Homo sapiens*. This is co-evolution as for previous *Homo sapiens*, with a difference in the relationship to the size of the group or community. *If the size of the group is too small, the difference specific to sapiens may not be perceptible.* Above a certain threshold, innovation increases. We can imagine a model in which the rate of innovation would be proportional to the number of possible interactions that increase, such as $n(n-1)/2$ with n equal to the number of individuals in the group. This model needs to be made more complex, because the capacity to transpose is also the capacity to transpose prohibitions. Negative experiences can cause lasting blockages. Stasis – the absence of innovation – in *Homo sapiens* can be long. Explosive innovation can also appear. *Homo sapiens* could resume evolutionary patterns according to the logic of punctuated equilibria, on tighter time scales.

Any new technique actually selects those in a population who know how to use it best. The piano is a recent invention and there are "geniuses" of the piano. Now that this instrument has spread in Asia, very great Asian pianists are appearing. It is the same with cycling, football, mathematics, computers, etc. It is as if, in a population, invisible technical abilities are selected by a new technique; in any case, they are highlighted. It is not the technique that creates these new capacities; they are there, latent, appearing by exaptation, and we cannot have the witnesses of an imaginary experimentation allowing us to know what would have become of those who show these abilities, which are not detectable, in the absence of the corresponding tool. How such capacities emerged in the initial absence of selection pressure by these new techniques is not in itself a problem, because it is the capacity for neuronal exaptation that was selected during the long process described above. Neurology should eventually enable us to understand, explain, and one day, anticipate. It remains certain that new techniques will emerge and that those who show themselves to be the experts will not be identifiable *ex ante*.

14

The Domestication of the Wolf: A Decisive Advantage?

14.1. The oldest domestication

The domestication of the dog is now described as a domestication of the wolf; it is another example of the problem of older than supposed achievements. For about 30 years ago, the domesticated wolf was assumed to have been domesticated a little before animals of agricultural interest, 14,000 years ago, on the basis of archeological surveys. This in itself posed difficult problems for the following reasons. First, the dog is present in all human cultures, including Africa, the South American Andes and Australia, where the dingo, a yellowish dog returned to the wild, is found (Smith and Litchfield 2009). However, humans arrived in Australia more than 40,000 years ago and were probably accompanied by dogs. *Ad hoc* hypotheses assumed that the dog arrived 5,000 years ago, or even 3,500 years ago, by Austronesian navigators, which would have allowed its spread from North-East India, Indochina, the Indonesian archipelago, the Philippines and New Guinea. This is typically the *ad hoc* hypothesis in solving an embarrassing question. For as we now know, domestication had taken place long before.

The morphological variability between dog breeds is exceptional, to the point of exceeding all known variability in a mammal. It is confirmed by high genetic variability (Vilà *et al.* 1999) and profound neurological changes (Frederick 2019). All this suggests a very long history of domestication, associated with subsequent interbreeding in human movements over the last 50,000 years (Druzhkova *et al.* 2013; Thalmann *et al.* 2013; Skoglund *et al.* 2015; Pires *et al.* 2019). In addition, the dog, in contact with humans, has undergone profound changes in terms of its cognitive, affective and dietary habits (Cohn 1997; Hare 2002; Kaminski 2004; Hare and Tomasello 2005; Axelsson *et al.* 2013; Losey *et al.* 2013). Even if the data are varied and sometimes contradictory (Perri 2016), it can be considered that a kind of theoretical convergence could be drawn.

Within a few years, Mietje Germonpré (2009, 2012) transformed the idea of a late domestication, estimated at 14,000 BP, by finding dog fossils on Aurignacian sites. Furthermore, crossbreeding seems to have taken place (Druzhkova *et al.* 2013). We can bet that we will go back beyond 35,000 BP for the first "domestications", followed by multiple intercrossings. For the genus *Canis,* despite its de facto morphological variability, is a genus in which almost all the species are interfertile, which may have been the case for the genus *Homo*. Molecular data, apart from confirming the astonishing variability of the dog, shows that the maximum proximity is with the grey wolf of the Middle East and takes the theoretical separation back more than 100,000 years (Vilà *et al.* 1997, 1999), but this may depend on the breeds of dog considered. It seems that the originality of the human-dog relationship needs to be reconsidered as a co-evolution rather than a domination of one by the other. The comparison of the history of the genera *Homo* and *Canis* is interesting for a general reflection on the relationship between human and dog. In the genus *Canis,* it is indeed the dog that succeeded, just as in the genus *Homo,* it is the modern human. Today's wild wolf is perhaps a specialization of the ancient *Canis,* which would be the ancestry of the wolf and the dog (Schleidt *et al.* 2003).

It is difficult to question the separation that appears clearest between dog and gray wolf in the Middle East, in molecular terms, in 135,000 BP (Vilà *et al.* 1999). Would there have been an *Out of Africa* (the penultimate) more than 100,000 years ago that would not have gone beyond the Middle East? Groups of *Homo sapiens* would have gone to the Middle East and encountered a herding animal with whom a "friendship" was established. So there are still many unknowns with a strong hypothesis: the initial domestication of the wolf is much earlier than assumed.

14.2. The co-evolution of humans and dogs

We could formulate the following scenario: the progressive adoption of the wolf took place during this penultimate trip out of Africa. This domestication led to a morphological transformation of the dog into several large groups. Even if the dog breeds were not stabilized and listed until quite late, around the 18th Century, archeological data suggest that large groups of dogs existed in antiquity (molossus, pointing dogs, racing dogs). For example, today there are more than 30 breeds of greyhounds whose origins range from Central Asia to Western Europe and Africa. Greyhound racing was already in common use during ancient Egypt. Turning a wolf into a greyhound does not take only a few centuries.

The dog-human co-evolution turns out to be long-lasting, so we don't see why, in such a close relationship, humans would have evolved less than dogs. Miguel Benasayag (2016) writes: "Thanks to this specialization of dogs, humans have been freed from this function [the sense of smell] and, in doing so, have gradually transformed their facial structure, particularly the nose and mouth [...] which has made possible the selection or development of organs that allow the production of more elaborate vocal sounds, up to articulated speech." Even if it goes a little too far, given that facial structure and speech had evolved long before, at the time of the emergence of *Homo neanderthalensis* and *Homo sapiens*, the image is indicative. Benasayag's idea is that the domestication of the dog would have been the ultimate stimulus for the liberation of speech, which is in line with our hypothesis. A human-dog co-evolution would have been added to the human-technical-society co-evolution. This means that dogs were already accompanying humans 70,000 years ago, which is consistent with the molecular data, and that dog-human co-evolution may have been the ultimate selection pressure leading to the human of the "cultural revolution" or "human revolution" (Chris Stringer 2012), which we will now call the "first cognitive revolution".

It might seem surprising that the reference to the dog in archeological excavations at sites dating from 50,000 to 15,000 BP is so weak. This confirms what we have already shown, that without a theory, it is difficult to observe correctly. It took only one archeologist, in five or six years, to set back the observation of dogs on archeological sites by more than 25,000 years. Since the data on dog domestication has been confirmed, researchers are discovering the importance of the dog in the last trip out of Africa, to 50,000 BP. It can therefore be assumed that from the groups that domesticated the gray wolf in the Middle East about 100,000 years ago, some came back in Africa. Believing that African populations remained isolated, stewing in Africa, while groups conquering the world made human history is also a belief that prevents us from grasping the complexity of exchanges between human groups.

There is almost no human group that does not have dogs. The latter is commonly called "man's best friend". On the last exit out of Africa, the human groups could have been accompanied by dogs, which meant that those who went on the penultimate trip came back with dogs or that dog exchanges took place. It's a bit as if the penultimate trip had been an exploration. In response to the treasure found – the wolf "friend" or "ally" – they brought it back to the land of their ancestors to the point of creating a revolution. This remains a hypothesis, because dog tracks in Paleolithic caves, whether footprints or representations in paintings, were rare or even absent. Scenes from everyday life were exceptional; the dog is, at the level of everyday life, almost never sanctified. Many people still eat dogs today.

Scattered across all continents, all groups arrived with dogs, which were often selected according to need (e.g. sled dogs as soon as they arrived in snowy areas). Crossbreeding was probably undertaken with other wolves and/or other canids, which opens the door to another understanding of the theories implying multiple possible domestication from the Siberian wolf, silver foxes, coyotes and European wolf. The dogs of the European Paleolithic seem to be the result of specific crossbreeding. Today, it is still possible to cross a dog and a wolf, and any species of the genus *Canis*. As a result, genealogies are likely to be complicated. As is the case, this reinforces the plausibility of the hypothesis: very early domestication, multiple later hybridizations, to the point that traces of the earliest domestication are difficult to detect.

The dog has been acclimatized to human life to the point of becoming omnivorous, which cannot be done in a few thousand years. It is assumed that human consumption of starch products in large quantities is fairly recent (less than 20,000 years); this may be questionable according to theoretical analysis. The use of fire led to the consumption of starch products, seeds or roots, that is well over 500,000 years ago. Let's wait for the analyses of molecular biologists and chemists. Dogs can consume starchy products whereas wolves are unable to do so, because from the beginning of domestication they had to be fed with the remains of cooked products. The dog has been under selection pressure in his role in the relationship with humans; this selection has been so strong that the categories of dogs correspond to specializations. To this day, the date of the mutations corresponds with the appearance of agriculture. This is an indication of the last selected mutations. In the initial bush-like evolution of the dog breeds, it is possible that many lineages have died out. And by crossbreeding, any gene can be transferred between populations of the genus *Canis*.

In retrospect, humans can be considered to have domesticated the dog by subjecting it to a typically technical selection pressure, initially oriented towards protection, hunting and entertainment. Since it seems reasonable to assume that this beginning of domestication took place long before the so-called "human revolution" or "cognitive revolution", it is reasonable to ask: what role did this domestication play in this revolution? We can hypothesize that it was the result of this return from the Middle East with this new companion, wolves tamed as dogs. It has transformed the balance of power. Groups accompanied by dogs became almost invincible. For any protection or hunting dog does not hesitate to sacrifice itself to save its master.

Associating an animal that was in fact a competitor for the same type of game and which, apart from the technical component, hunted in the same way, in a group, with a highly elaborate strategy, was certainly an additional experience founding otherness. So much so that the question is who adopted whom: was it the wolf that chose the human[1]? We agree with the reflections of Schleidt *et al.* (2003). We can imagine that the cub's behavior was so appreciated that the selection by humans favored the maintenance of these characteristics in adulthood, in addition to a search for adaptation to specific tasks. This means that humans imposed on the wolf what their distant ancestors had imposed on their children, and probably with less care. By imposing this companion on themselves, they evolved with it. The dog appeared in the human technical environment. It would suffice that the human groups' domestication of the wolf has gradually appropriated different individuals, open to the relationship with humans, less defensive, so that, with moderate selection, a new population of different wolves begins to associate with humans.

Humans have adopted a selection system for dogs similar to the one they have used for millions of years, on themselves. With three obvious specific components: a wolf is not a superior primate; the relationship established is a relationship of dominance; the direct use of techniques by the wolf was

1 Research has questioned unilateral domestication; there is at least one case of cooperation between monkeys and wolves in Ethiopia (Venkataraman *et al.* 2015).

unimaginable, even if it could pull loads. The rapid evolution of the wolf into a dog inspired Darwin in his analysis of breeding. We cannot but be struck by the morphological, behavioral and physiological impact that made the dog an undeniable ally in the conquest of the world. It is known that molossus are war dogs that were used by the Romans in the Punic Wars and by the Spaniards during their conquest of the Americas. It cannot be said that they were widely publicized. No land animal, no matter how big, powerful, aggressive, and living in a group, could now resist a group of human hunters accompanied by dogs ready to sacrifice themselves for their master. The dog, selected for its barking ability (neotenization), which existed but was less developed in cubs, became an additional night-time protection. Night-time social life around a fire could continue to develop under the protection of dogs; the nights could be even more serene.

The weakness of the analysis of domestication, which is a process dependent on technique, since the selection of the dog was carried out in a human technical environment, is similar to the weakness of the reflections on technique in human life. The domestication of the dog was the first reception of another in the human environment. It seems that certain breeds were selected for the protection of children, so striking are the friendship and the delicacy of the behavior of certain breeds with children (Newfoundland, Labrador, Golden Retriever, St. Bernard, etc.). We have the right to wonder whether, without dogs, agriculture and breeding would have been possible. With dogs, crops and herds could be protected.

14.3. The strength of the association between humans and dogs

At the meeting of *Homo sapiens* and *Homo neanderthalensis*, between 40,000 and 30,000 BP, if population ratios played in favor of *Homo sapiens*, it is highly likely that the presence of dogs played a role. With such a companion, he was able to take better advantage of different game and had easier access to a varied diet. Moreover, the techniques used by the *sapiens* seem more differentiated and adaptable. It only took a differential in population growth to overwhelm the Neanderthals (Cohen 1999, 2007; Maureille 2008; Hublin 2011). Hybridizations between the two "species" have taken place, even if there have been confrontations, as between any group. On a simple differential of 10% per generation, lasting between 5,000 and 10,000 years, the differential growth changed the population ratios by a factor much more than 1,000. The Neanderthal population was small, probably less than 200,000 individuals, and seemed to have been stable for several millennia, over an immense territory. The world human population was already estimated at a few million at the beginning of the Neolithic period. The major hypothesis of hybridization and demographic overflow seems the most likely.

It can also be said that from the "human revolution" from 100,000 to 50,000 BP, human populations were growing, despite various premature deaths, which would make it possible to understand permanent migration on the one hand, and regulation by different cultural practices on the other. The technical explanation for the higher fertility of the *Sapiens*, compared to pre-human ancestors, is probably daring; however, if it is confirmed that there was a bottleneck shortly before that time, and the very fact of the expansion that followed just afterwards means that the potential for population growth was high, whether in terms of natural fertility or general child welfare.

The thousand-fold increase in human populations in about 40,000 years, as the facts show, meant an increase of less than 3% per generation. For despite a fertility potential greater than that of a primate, since women are fertile every month and for longer periods, the nomadic lifestyle led to a very close relationship between the mother and her children for three years. The mother carried her child (see Timothy Taylor's theory) and breastfed it continuously, delaying the resumption of the menstrual cycle after birth. The sling, while it freed the group, allowing it to move more freely, was probably a birth regulator which selected a lengthening of the duration of fertility. And dogs were there to carry or pull loads.

At the time when *Homo sapiens* became the "master" of a species already belonging to the "masters", large predators, at the top of the food pyramid, it is conceivable that this new condition was favorable demographically speaking. Big and small game became even more accessible, the protection of children was assured, and the peace of the night times was guaranteed. One still wonders what happened during this period that opened up a new techno-cultural transformation. Probably three interacting events: the domestication of the dog, the finalization of the mastery of a doubly articulated language and a demographic growth that accentuated the capacity for innovation in stages. Directing an animal may not require language; it does, however, require precise understanding, foresight, behavioral understanding, the ability to listen to a different body language, and the creation of emotional bonds that transcend the species. The she-wolf raises her young in burrows and it seems that humans only became interested in caves after meeting the wolf. To have a companion of another species, after having been able to invent a companionship with our own tools, requires a unique psychic flexibility. In both cases, a sufficient number of generations was needed, both to capitalize on knowledge and to stabilize the populations of the groups, which were probably increased.

We have an additional fact to support the argument that dog domestication was very early. It comes from linguistics. Among the twenty-seven global roots selected by Merritt Ruhlen (1994, 1996), which make it possible to conceive the existence of a mother tongue of all current languages, just as the original Eve was searched for by the mitochondria and the original Adam on the basis of the genes of the Y chromosome, we find the word that names the dog. Before the differentiation of languages throughout the world, as this term is found in African languages, the word designating the dog existed to the point of being a collective identifiable root (Kuan). In order to designate the dog, it had to be already present in the human world, which is the origin of today's languages, in Africa.

It is quite possible that even today we still underestimate the impact of this so particular "friendship" between man and dog in the constitution of the human. He is now learning to live with a living being of a different species, with different needs, and yet he brings him into the human clan. This association in fact diminishes the difference between humans and the living environment; other living beings can be welcomed; the possibility of modifying the environment to make it conducive to others than just human life has just emerged. According to the writings of primatologists, a chimpanzee integrated into a human group (that of primatology researchers) will rank in the world of humans and not chimpanzees. Similarly, the dog is now part of the human group and not the wolf group.

15

Reforming Our Thinking About Humans?

15.1. The human characteristic: a search without a future?

Neoteny, bipedalism, the growth of the cephalic index, the use of tools, a taste for games, learning or language are not absolutely specific to humans; they are always present in other living beings at different stages of development. Similarly, the human brain functions from the same neurological and biochemical processes that are present in all mammals. It is "only" much larger and more differentiated, with intermodal functions becoming more and more important, but some modules have also increased in size, sometimes with specific functions. The evolutionary process from which it originates can be thought of according to the Darwinian scheme, provided that we add the role of technique, which often results in the inversion of selected characteristics.

Is it possible to look for a human characteristic which is unique and enables him to assert his absolute originality in the living world? Since Descola (2013) and the end of the separation of nature and culture, it was suspected that it was not. Nevertheless, a specific combination, which will have to be analyzed more precisely, has led to this speaking living being, which can feel and contemplate the beauty of the world, imitate it, represent it, create something new, sing, dance, play roles, analyze the reasons and necessities, unfold a complex history, establish links and understand relationships. Drawing a picture of the world on a surface, telling the passing of time and imagining the birth of the world, transmitting and sublimating emotions, building collective rules and laws and adapting to them, anticipating independently of one's needs of the moment, loving and getting angry, destroying and creating, and integrating all these capacities into action by building new worlds. It is this set of qualities, developed differently depending on the individual, which seems extraordinary to us and leads us to want to separate the human from other living beings and to look for our own, especially since action has always been the first for every living being, and Man, without being original in this, only pursues the integration of new capacities in action and through action. These so-called human qualities are collective. No human being develops them all to their possible climax. It takes a group of sufficient size and special conditions for them to be revealed and to flourish. Human evolution, after the emergence of *Homo sapiens*, will show us this to a large extent.

Laughter is not a human trait, nor is artistic or political meaning, nor imitation, nor singing, nor the ability to emit a succession of sounds transmitting information. However, we intuitively feel that there is something false in this research that wants us to admit that there is nothing in humans that is different from the rest of living beings. Pascal Jouxtel (2005) admits that he enjoys destroying any claim that there is something original about the human being. However, it is doubtful that individuals of another species entertain themselves in this way. Francis Kaplan (2001) and Jean-Louis Dessalles (2007) have clearly shown that the doubly articulated and representative language, which tends to universalize what is said, is unique to the human species. To this can be added a huge list of human achievements that have no equivalent and for which there is no guarantee that it comes from language. We have seen it with the tools: intelligence and perceptive thought developed long before language, starting with technique. This apparent contradiction must be understood here, as well as the often exposed feeling of incompleteness. The human being as an unfinished being, always on the move, is not something new either; it is in a way the very own world of the living considered as a whole. The human being reveals it more intensely. To want to spread, to multiply, to hit the limits of existence, to push them back, to transform our environment, is to be a living being. But, as far as the human being is concerned, we do not know its limits; they seem to go beyond the field of the living world itself and this problem becomes a field of study in itself.

Morality doesn't seem to be specific to humans. Primatologists defend the existence of a moral sense in the bonobo or chimpanzee. It was believed that in evil, man was fundamentally different. Even then, the preconceptions have been contradicted. In many species, fights between males for the conquest of a female can lead to the death of one of the suitors. Worse, the winner can kill all the cubs of the females he covets. Destructive fights, real wars, between groups have been shown in chimpanzees. The so-called natural law to be applied in the human world would be catastrophic if we took this claim literally. It would be the individual struggle for life and the proliferation of each group against each group where all blows would be allowed.

How can we get out, if it is possible, of this astonishing contradiction that leads us both to the conclusion that the human being is absolutely original and to the conclusion that he is only pursuing a tendency that is specific to life? We will take up the account previously presenting and, using it as a basis, analyze the source of this contradiction.

15.2. The major innovation in body techniques

Edgar Morin remarks that things like funfairs, circus games and magic tricks are universal. He takes up the analyses of Johan Huizinga (2016) and Roger Caillois (2001). As if, there again, a characteristic of the human being was revealed. In circus games and magic tricks, what attracts people is the confrontation with technical and emotional difficulties and the demonstration of their resolution by doing so. Wonder comes from the exposure of abilities beyond those that we can feel for ourselves. With this spontaneous question: How is this possible? How indeed are the unbelievable physical exploits that some people know, possible? How is it possible to tame ferocious animals that are capable of killing a human in a single blow? How is it possible to perform elusive magic tricks? How are some people able to compose and recite texts that bewitch and transport to imaginary places? And others able to "speak" by whistling? Because, since the earliest known antiquity, competitive games have existed in all cultures in all fields.

Attention to our human world reveals that this "amazement" can be much more general. How is it possible to send humans into space and to the Moon? How were airplanes carrying hundreds of passengers at 900 km/h made possible? How could the properties of electrons be described? How can these "superhuman cities" exist? How can a symphony be conceived and written? How were these cathedrals built during the Middle Ages? How did the ancient Egyptians build such pyramids? How could megaliths be erected, more than 6,000 years ago? How was pictorial art mastered as early as the

Aurignacian (Chauvet cave)? How were human groups able to cross the last ice age in Europe and Asia when their ancestors came from warm regions and their young require care, attention and protection? How did the first vertical primates survive against various predators they could not run from? The list is a long one of possible wonders, provided that we retain the ability to do so, in the face of the achievements of humans and their ancestors. The one that takes place in front of circus exploits speaks for itself, because every human has been to the circus (or the equivalent).

In the latter case, we immediately see the technical mastery of the body which fascinates and leads us to raise questions, because we feel that it is an ancient and profound expression of the human, as if we had partly lost it. This is the case with the discovery of the exploits of fakirs or shamans. How do they control their bodies at this point? Or in an egocentric approach: why can't I, as a human being, do it? The human who fascinates us, who impresses us because we know how to be part of it, poses a problem for us, because we are most often forced to recognize our inability in this precise place where he expresses a form of magnificence. Pushing a human quality to its excellence is achieved only by a small minority of humans.

Yes, there are composers and musicians, architects, painters, writers, choreographers, scientists, engineers, athletes, strategists, philosophers and theologians, craftsmen, sculptors and stonemasons, blacksmiths, potters, etc., so many diverse talents whose list is immense, scattered throughout human populations. Anyone who reflects and considers himself part of this community, who recognizes this greatness of the human being, is astonished that it is carried, apparently, by a relatively small number, whose fame extends to all. He asks himself: how is this possible? How do you do it? Why is it so human and yet exceptional? What is this human world that appears so cooperative, whose remarkable capacities are shown in a small number of individuals, which exist collectively in and by that minority, and whose provenance is both universal, distributed among all, and at the same time limited? It is as if there is a human potential, fragile, capable of flourishing if the conditions are favorable, but which, generation after generation, remains a little stingy in showing itself fully. What we apparently see is crime, violence, disease, suffering, etc., but it's not the same for everyone.

The bodily techniques are multiform; they can be learned by anyone. Those who master certain aspects of them at best are admired as exceptional as they reveal what emerges from this collective human, this ability to use the capacities of the human body for actions that require, most often, but not always, a mastery of external tools. This is the case of verticality, of this posture from which, according to my analysis, comes this incessant and protean development of the techniques used, and which has oriented the future human; verticality learned by all and given to all, although, there again, some people use it more excellently than others. Verticality is both commonplace, because all humans stand vertically, so difficult since many cultures insist on mastering it, that some activities require a level of excellence reserved for an elite, while low back pain and other spinal problems are the lot of all.

The verticality that leads to the human would have appeared a long time ago, between ten and five million years ago. It must have been in its infancy, hesitant, lost and taken up again several times, and, as is often the case with evolutionary processes, in a bush-like manner, with certain lines dying out. Thus, verticality gave rise to many different lineages. There are still traces of this in chimpanzees and bonobos. In fact, it is such a strong tendency in great apes that when they crawl on all fours, they put their hands upside down, as if it were just to rest. Sitting or squatting in a partially upright posture is common among apes. It can be found in rodents or bears. The question is therefore not the originality of the upright posture, it is the establishment of a socially constructed tension, learned and transmitted, which forces us to stand upright permanently and which is selectable; that is, it has been maintained for a long time over many generations and is valued. This pressure for verticalization can be thought of as a collective pressure for collective integration, which can be described as domestication, because it is a process that has taken place in an already socialized species according to a socialized process. When a typically human process is considered natural (i.e. it is naturalized), it stops being questioned and is taken as a fact. As soon as it is considered as human, as unnatural, it enters into the questioning of what makes us human. We see it today in food. The contemporary questioning of the naturalness of human food is a real upheaval. We see the flowering of so-called Paleolithic foods or vegetarian, vegan foods. Will pre-Columbian, Tibetan and Australian Aboriginal diets flourish? Taoist? The same is true of verticalization, which we consider to be a characteristic of human beings and an evolutionary motor from the very beginning. It can then be questioned, and we discover its importance in many cultures.

We could believe that, after more than four million years of selection pressure, and an impressive morphological and musculoskeletal adaptation, this verticalization is definitively won and biologically integrated. Personal experience and observation show everyone that it is a permanent "combat". Comic strips and cinema insist on the difficulty of adolescents to acquire it and pejorative terms are not lacking to say that "righteousness", "desire" – or "energy" – and verticality are directly related.

The great civilizations that have developed methods of meditation all insist on posture. Whether it is Hinduism, Buddhism, Taoism, Chan, Zen, etc., the insistence is strong on two points: the vertical posture, even when seated, and the mastery of breathing. These are two fundamental bodily techniques, on which we insisted at length, the first learned during the bush-like evolution of the *Australopithecus* period, and the second during another bush-like evolution which produced *Homo erectus* and then *Homo sapiens*. Jacques de Panafieu (1979) rediscovers this ancestral knowledge in what he calls "intensive illumination" and discovers that, in the intensity of the search for themself, the participants in his seminars gradually straighten up, without having to ask them. Everything happens as if, in order to become fully human, and to discover ourself, this vertical tension was necessary. Asking the question "who am I", which cannot lead to a rational and objective answer, yet leads to straightening up and controlling our breathing. Yoga and the martial arts associated with Taoism have formalized precisely this requirement. The circulation of energy, breath or chi is thought according to a vertical logic. There is hardly a culture that has not symbolized the power of the vertical by systematically putting the divine attractive or "principle" at the top, and the despicable or to be rejected at the bottom. Language is full of metaphors that privilege the top and front over the bottom and back. Symbolically standing is being free and responding to a call; submission is shown by bowing, kneeling, squatting, or even lying down. The cultures of Middle Eastern monotheism do not seem to be explicitly concerned with the inner workings of verticalization; on the other hand, its symbolization is everywhere, in images, in art, in architecture.

Verticalization is still a learning process, again now; there is little doubt about that. How did it establish itself? I proposed a story, knowing that there are many other possibilities. Which could be the most credible? Undoubtedly those that involve collective processes, since this is generational learning. Can there be an inner demand outside of a call from the relationship and collective? Everyone is born into a network of relationships and is built through interaction, not just humans, but mammals, and of course monkeys, and among them great apes. If verticalization accentuates these interactions, it also comes from them. If chimpanzees have not gone further in the production of tools, it is probably due to the lack of a process of collective and individual investment in body technique. In the lineage of the hominins, our hypothesis is that verticalization has been the initiating process; it is not by chance that some have proposed an aquatic matrix, based on the absence of fur: it could have been swimming... So long as there is no concentration nor effort exerted on the body itself, a tool technique requiring double attention does not seem possible. The practice of swimming is different from the bodily activities that push toward verticality; in any case, still today, most of the techniques called "personal development" promote and value verticality and affirm an interaction between self-awareness and verticalization. The horizontality of swimming does not lead to the search for equipped activity, on the contrary, it leads to a regression. The hands are not free because they are necessary for swimming. It is very recently that the tool has been introduced. This could have an important meaning.

15.3. Technique and the game: a fundamental intertwining factor

The name *Homo sapiens* according to Edgar Morin could be changed to *Homo sapiens-demens*. History presents a good case for this; the same individuals, or communities, or even nations, reveal both human wisdom and folly, and it would be easy to show how pure wisdom is folly, and in some cases that folly is a form of wisdom (Foucault 1988). The term *Homo faber* is closer to what we have sought to show. It would be more precise to speak of *Homo technicus*, provided we take the term "technique" in the sense we have defined it, or *Homo docens,* according to Gärdenfors and Högberg, but this is more limited.

Johan Huizinga's proposal (2016) in his seminal book *Homo ludens* had a broad ethnological and anthropological filiation, because play and the sacred are inextricably linked. From the beginning, we suggested that the playful function should be present at the foundation of the technique. The definition of play, proposed by Johan Huizinga, inscribes it in the technique, since all play integrates our definition of the technique. The game can be considered as a framework within which the technique is inscribed or as a foundation from which the technique can be put in place. In the game, skill within agreed rules is indeed a relationship to learning. As for the delight and enthusiasm that the game provokes, it is not without relation to the pleasure of success, the pleasure of know-how and the pleasure of the work accomplished. The fact is that technique, like play, requires an action that is situated, that is to say limited in time, space and context.

As there is no *a priori* specification of utility in the technique, we deduce that the best way to learn to use a technique is to play it, in a time, place and situation where the stake is not precisely survival. This is what appears in the use of play in the young of mammals and birds. The more they are evolved, in the cerebral sense, the more they play. They play games specific to their species and thus learn the skills they will need as adults. It is known that in the most "evolved" species, some games are maintained in adulthood. The documentary *Bonobo* shows this to a large extent, since even in a tense situation such as contact with a snake, the playful aspect remains present.

Johan Huizinga suggests that the game is part of any technical practice, becoming both the driving force and the consequence. We play to learn, and we play for the pleasure of showing our know-how. We can play to ward off fate, define new "rules of the game" to avoid deadly confrontations. We play in a defined space, during a defined time, and according to rules that are not those of everyday life. When you play, you are out of the space-time of everyday life. The strongest experiences of childhood are rooted in this space-time out of everyday life.

Technique and play are two complementary ways of understanding the role of technique in human development. Technique can be understood as child's play that ends up taking up so much space because it offers new ways of satisfying adult needs, such as a taste for power. Play, as the mastery of a technique, becomes a serious game (in English in the original text).

We can consider that all techniques are also games and it appears that if technicality reveals itself as a foundation of the human, it is probably through the game that it never ceases to impose itself, because not only has the game probably been at the foundation of the technical process that leads to the human, but it flourishes all the more as the human reveals himself more. The neotenization of the human is the extension to the whole human world of what seemed to be reserved for the child as a learner and discoverer. Inventing is indeed child's play, and the difficulty of invention comes mainly from the difficulty of keeping the spirit of childhood present in adulthood. It will remain to understand the relationship between technique, play and the sacred and its evolution from the beginning to our times.

15.4. New accounts of the emergence of culture

What can be the effect, in the long run, of such a posture, whose inheritance accumulates and is amplified over time? Let's think with Darwin: what is the effect of a "small" trend, probabilistic, selectable, which lasts from generation to generation for at least seven million years? It must progressively transform the living beings that manifest it in an incessant succession of generations; moreover, we must take into account the systemic aspect of the interactions between the pre-human and his tools, as well as the mediating role of the tool in the relationship with nature. Darwin (2019, 1882) has repeatedly shown that a small cumulative variation over time can end up creating large macroscopic phenomena. If this small addendum is established in a systemic relationship with positive feedback, the effect will be even more pronounced.

The first transformation manifested itself in three main stages until the emergence of *Homo sapiens*, ready for visible and accelerated cultural evolution, that is between 70,000 and 50,000 BP. Verticality, as an internal tension and social control, can be thought of, afterwards, as a takeover of the group's power over the "nature" through the construction of a "social" more protective organization; this collective phenomenon, which is maintained and amplified, creates a terrible selection pressure which, at least for the lineages that maintain this pressure, is more effective than any return to a specializing naturalization that ends up being extinguished. Verticality, and consequently the opening to a wider world, leads to widening the horizon and, in concrete terms, to widening the field of what is consumable; each selected modification that widens this field increases the selection pressure in the sense of technicality. Neither the worst of predators nor even terrifying natural phenomena frighten the verticalized group any longer, which conquers and acts together (except in the cinema, where one goes along with being afraid). Bodily learning is a flame that grows from generation to generation or is extinguished.

The footsteps marked in stone at Laetoli and dated at 3.5 million BP give us food for thought. The biped monkey seems to be no longer afraid of anything, or, if he is afraid, he nevertheless faces the elements and the unknown, walking barefoot in the volcanic dust. It is not his weakness that will lead him to create a culture; it comes from pre-verticalization protocols. It intensifies as a new emergence from this tension to straighten up and look away, a tension that is transmitted and continues, from generation to generation, in a body more controlled by an increasingly complex brain. Culture develops with the technical protection that has made it possible to respond bodily to this demand for verticality by increasing the brain's potential to the point of provoking new capacities of representation. Finally, technique is the underpinning of culture, which is the meaning constructed from technique. Little by little detached from a so-called natural world, humans are entering the "magic" world of representations, of the search for reasons and links, of the integration of duration, that of language, of culture, with blurred and fluctuating, changing and manipulable meanings. The complexity of the brain shows the complexity of its relations with the body, its organs and all that "exosomatic" peculiar to humans: the artificial world it produces and that transforms it.

Its apparent fragility flourishes in technical protection, as the hermit crab is protected in its shell, but also in collective, social and cultural protection. In the hominin, this shell is transparent, moving, in permanent construction, both abstract and concrete, and all its senses make it perceive the immensity and variability of its world. In order to find the reasons that make him leave the world and yet see souls similar to his own everywhere, he imagines a magical protection of his group by what his group came from, in ancient times, and which he lives as a living reality. The process can only continue unless he alienates himself from another power that takes his freedom, which becomes possible as soon as a human group goes far beyond Dunbar's number and instituting political strategies become necessary for the group's survival. But this is another and the next story.

From Peter Sloterdijk's analysis, we can imagine the beginning of insulation, of isolation of the group, corresponding to the amplification of the social space created by the requirement of verticalization, separating it from the environment which has become a space of conquest. The distancing from the environment, in the sense of von Uexküll, whom he calls the "suppression of the bodies", leads to the progressive creation of a "world", his own, which is a collective world, a milieu, and which he will incorporate little by little, after a few hundred thousand generations, according to the cerebral development allowing individuals to internalize this world. He will have gone through several phases of neotenization: that of stable verticalization, that of the growth of the cerebral neocortex, that of the lengthening of childhood, that of the preservation of the spirit of childhood even in adulthood, which exapt neotenization itself. "It" is, of course, a metaphor, for the hundreds of thousands of generations have continued, and while we can, today, acquire an awareness of evolution towards the human and the transgenerational adventure it entails, this concerns, for the moment, only a small number of us. The only perceptible developments are the technical, economic, social and modern breakthroughs.

From the very beginning of verticalization, "it" gradually learns to transfer its new knowledge. After the throwing of stones, it is the use of broken stones, then the knapping of stones, the use of piles, the construction of habitats, baby carriers, the use of fire, the enrichment of hunting methods, etc. The narrow clearing opens and lights up. Everything he learns can be exapted, everything he understands becomes a source of learning to understand. By looking for reasons for everything, he looks for reasons to respond to the astonishment of his own existence; can one be conscious of oneself and of the world, can one change the world and at the same time be part of the world? The whole journey towards the human being is set in motion, certainly full of traps, happy or unfortunate accidents, with risks of failure – and there have been many of them. Finally, we recognize, in this beginning, with the benefit of hindsight, what made the human being.

How does this story integrate the requirements of François Sigaut? Certainly, the sharing of attention is required for a bodily technique and the vertical posture. It is indeed a movement, a founding physical position, but it would be naive to believe that this sharing of attention emerged suddenly. The first groups of primates probably evolved in this direction in an approximate way, by trial and error, perhaps by playing, because it is the little ones who are concerned in the first place. There is no reason why there should be no individual variability in this ability. As a result, a collective selection pressure had to be put in place (see Peter Sloterdijk's insulation and the retouching of the neotenic narrative).

The pleasure of success is not unique to humans, but it is more intense in them. It is doubtful that it can emerge without the gaze of another, even if there is, in the pleasure of success, the mere fact of succeeding; there are graduations. By positing that the evolution towards verticalization is learned, marked by the imprint of the relational, success can only lead to the sharing of the experience and its amplification. Two aspects can be added to

François Sigaut's analysis here. In technical activity there is a singular pleasure, that of the inner perception of mastery in the activity. This can only be added to success, but amplifies the desire to begin again, to take it up again, as Kierkegaard calls it; it is a form of communion with the body and the objects used that makes us perceive the fluid interaction that goes from the psyche to nature, passing through the body, the tools and instruments. Technical mastery often leads to the desire to rediscover the corresponding, almost magical inner perceptions. The second aspect, certainly after the pleasure of success, is the taste for a "good work". Without even needing collective support, everyone perceives when it is a beautiful work, or on the contrary a "dirty" work, a "sloppy" or "botched" work. Making a beautiful work is a great pleasure, even if it is not shared. We could add, according to Huizinga, the agonal aspect, that is to say the taste for challenge and competition, associated with cheerfulness and enthusiasm.

Chimpanzees sometimes hunt in groups; the learned verticalization leads to more sharing, more cooperation. It is conceivable that gradual verticalization leads little by little to a selection of a more intense sharing of experience, indirectly, through collective action. We have already seen that, here too, insofar as adult verticalization corresponds to a lengthening of childhood quality, it is a *de facto* experience of exchange between the sexes. Thus, we can conclude that the exchange between the sexes and, consequently, less sexual differentiation has had to follow the process of verticalization. The archaeological archives show this dedifferentiation. Let us keep in mind that this account suggests that gender differentiation in the use of techniques must have been much more restricted during the journey to *Homo sapiens* than it will be afterwards.

All this leads to a metamorphosis in social organization and the beginning of monogamy. We can imagine, over a long period of time, from the first vertical primates, multiple activities shared between males and females. An important step is the male's interest in his children. Mating face to face, as in bonobos, because of verticality, is the opening to a form of equality in the relationship. Within this framework, all the premises of a human society are present with a mixture of domination-submission relations and egalitarian exchanges.

15.5. The influence of techniques on evolutionary processes

Language is a complex body technique; speaking can be learned, it is a skill (Carus 1893). Everything is already built or under construction, at birth, according to an ontogenetic process, in order to be able to learn; even if it is in the interaction that the central nervous system becomes more complex and that neural networks are set up, in order to be able to speak, there are morphological components that are only ready around the age of two: movement of the hyoid bone, release of the vocal cords, etc. (descent movement).

Language learning is rarely quick; it often takes two years to acquire it in a basic form, and then mastery of the language takes many years and may depend on parental mastery. Learning to speak is done both by copying and through a long process of trial and error. Without someone to talk to, the child would not speak. Handling objects is a long learning process. If the variability of talents is shown from a very young age, the fact remains that the first years of life are at the same time a physical development and a cerebral maturation which are intertwined in the acquisition of various body techniques, tool techniques (most often in the form of objects) and of course language. Language itself is constructed from body techniques common to singing: control of breathing, emission of sounds, articulation of phonemes using the tongue, the lips, the palate and its soft part, the teeth.

All body techniques are based on unconscious processes associated with conscious commands. In fact, we do not voluntarily control all the muscles that could be controlled and most of the time we let autopilot take over. Our brain, monstrous as it is, does not seem capable of exercising complete coordinated control of all striated muscles, just as it is incapable of holding together two different representations of the same design. This is what many "optical illusions", the duck/rabbit of psychologist Joseph Jastrow proposed in 1899 or the girl/old lady, or the profiles of Queen Elizabeth II and Prince Philip surrounding a vase tell us[1]. These are not illusions; it is a way of showing that we cannot see two different images of the same reality at the same time. In the same way, we cannot sustain an intense conversation by having another such intense manual activity. Our brain is such that none of us are capable of developing it in all directions: we have to choose. It is both very large and too small. It can only express its potential on a collective

1 You can find all these "optical illusions" on the Internet.

scale, in a society that offers different ranges of activities. This development will be, for each individual, partial; it will not concern the full range of its possibilities. It is conceivable that, at each phase of its development, groups have grown larger, as the sharing of new tasks and the corresponding learning processes required more people. The human being is, in fact and in principle, collective; he is shared; he is built in and through "us". It is accomplished individually by specialization.

The search for the possible succession of the different acquisitions requires entering into the logic of an evolutionary thought, that is based on the Darwinian algorithm of "variation-selection". We will take as an example a type of non-evolutionary reasoning applied to this process and show why it leads to deadlock.

Jean-François Dortier (2004) presents the analysis of the possible links between the birth of the tool and the birth of language, and explains, based on scientific controversies, that four figures are presented and discussed:

– language and technique have developed as two independent factors;

– language is the "driving force" of the appearance of technical intelligence;

– technical intelligence (tool) is the cause of the appearance of language;

– language and technique are the expression of a more fundamental skill that has conditioned their development.

At first glance, this seems logical: we use the axiom of the excluded middle and there are only four combinations of the two parameters taken into account; except that we are not studying a logical, static problem, but an evolutionary process. Hegel understood, more than two centuries ago, that logic, using the axiom of the excluded middle, is not adapted to the analysis of change. In an evolutionary process, if we have to think about evolution, we have to be able to think about chance and contingency, which is a difficult exercise for a logician or metaphysicist. This is the problem with these four figures; they are not in the same set of possibilities.

The first case suggests that if language and technique are independent, they should not interact, which is contrary to the facts. Chance has been described since Antoine Cournot (1843) as the meeting of independent series. The answer to this scenario is therefore "yes and no" according to what is meant by independence. That they show some form of reciprocal independence, to be specified, is evident; that they are neither interacting nor in a conditional relationship is contrary to the facts.

The second case directly states a kind of absurdity: how can language be a cause for the appearance of a practice? This is Aristotle's efficient cause, the cause of scientists. The added adjective (motor) further obscures the point. So the answer is "no". On the other hand, if we want to say that the existence of language could be a necessary favorable condition, under a specific selection pressure, for the appearance of technical intelligence, then we say that *Australopithecus* could speak. However, it seems that Broca's cerebral area is too small for this to be credible, not to mention the position of the hyoid bone and the whole respiratory system, which do not seem to be ready. The answer would be "no", under the conditions, even though language has had a role in technical evolution.

The third case states an absurdity similar to the previous one, and therefore the immediate answer is "no". If one wants to say that technical intelligence is a necessary but not sufficient condition, the answer can be "yes", especially since language is a technique. As we have seen, this intelligence is, among other things, the intelligence of reasons. If chimpanzees can learn a strictly descriptive protolanguage, which is in fact a pre-language, none of them seems to have access to the expression of conditions or reasons, nor to speak about the past or the future or to express the conditional. As shown above, they can speak, but they have nothing to say, implying: to say is to expose motives, reasons, links, duties; it is to address another. This is the case with the first ramblings of a two-year-old child. Technical intelligence could be a selective condition of language possibilities, but not of his final development. Let us say that technical intelligence can be conceived as a first selective condition of the basic possibilities of language, which, moreover, is a technique.

In the fourth case, the register is changed; there is no longer a cause, only conditions. There would be a fundamental condition common to technique and speech. Here the answer is clear: "yes", there is at least one common condition: the possibility of developing bodily techniques, because without them there is neither tool technique nor mental language/technique. They are not necessarily the same and the question here is whether there is a common source for all these bodily techniques. Our hypothesis states that from verticalization as a psychosocial construct, that is as a bodily technique, there is the possibility, by transposition/exaptation, of bringing about other bodily techniques, then equipped with positive feedback, on the condition that there is a valid variation-selection binomial.

Nevertheless, this last figure is of such a generality that it does not allow us to understand it. It would be like talking about the dormant virtue of opium: the creationist temptation is perceptible. There would exist, or would have existed, a fundamental aptitude which would have produced the technique and the language, that is the aptitude to become human. Language is a body technique: it mobilizes numerous and varied muscles, as well as a complex musculo-osseous and/or cartilaginous structure, and requires, in order to be mastered, complex innervation and a centralized processing system which is capable of providing interpretations, research into reasons, and various questioning. The equipped techniques and their bodily counterparts can be classified in order of increasing complexity, or better still of increasing efficiency, which corresponds to a chronological order in the lineages that lead to the human.

We must therefore return to the question: is there a level of technical complexity that would require an adapted language in order to be transmitted effectively? This is tantamount to implying that language has emerged as the technique is complexified; one need only look for that moment. This leads to the reverse question: can language, whatever the reasons for its appearance, influence the technique and its use? These two questions lead to analyses of reciprocal interactions and conditionalities, which are tantamount to posing the demands of evolutionary thinking.

Language, as it works, requires such structural and functional complexity that one suspects that it took a long time. The first techniques were already present in Early Oldowayan[2], more than three million years ago. It is therefore "evident" that the technique appeared long before language; it is no longer a matter of theorizing it, but of understanding it. How can we speak without the breath having mastered at least a minimum of one? How can we speak if the necessity (or the need) to say the reasons, the motives, the links, the becoming, is absent? It also requires body morphological conditions, muscular function, the ability to exapt body techniques, an interest in reasons and connections, and a selective collective pressure to promote exchanges.

2 Some people now speak of the Lomekwien (–3.3 million years), clearly identified as predating *Homo habilis* and therefore belonging to an *Australopithecus*.

The Oldowayan or Lomekwian techniques, though so ancient, are of an astonishing complexity that a chimpanzee is incapable of mastering them. It requires both a knowledge of the reactions of matter and a dexterity of gesture. This precision of gesture presupposes a development of manual body techniques. The change in the conditions of existence acquired from this technicality – which itself evolves because it creates the conditions for selecting its own improvement – will have to last a long time for it to induce a biological evolution, albeit contingent, from *Australopithecus* to *Homo habilis*, or *ergaster*, or the other more or less contemporary ones. Indeed, more than one and a half million years were necessary for the profound upheaval of the body and the brain, and for the emergence of a more advanced technique. When we know how to sculpt stones intelligently, we know or will know how to do many other things, exaptation permitting.

Searching for words, expressions, formulations, sequences to expose our thoughts is a common experience. "What is well conceived is clearly stated" is not universal. When it is true, it is because everything has been thought in terms of language; this is almost never the case when it is an invention, a discovery, an opening to the unknown. Einstein exposes it with remarkable humility. This argument alone suffices to understand that technique is anterior, but that from a certain level of complexity, its maintenance and evolution requires a representative language, and not simply a communication system.

We thus have an intellectual "tool" that can put the succession of processes back in order: evolutionary thinking from which the integration of the influence of tools on changes in selection conditions is achieved. The use of tools, by changing the conditions of existence, accentuates, displaces or modifies the pressure of selection, and can consequently play a major role in the orientation of evolution itself.

Climate change, in time and space, is a formidable selection system; it favors, each time, the collectives that have created their own robust environment. In fact, the groups that are least dependent on the specificities of a context will be those that will cross the obstacles while sustaining the least damage. Strong variations in the environment favor those who know how to use the opportunities available to them; they cause a progressive elimination of those who are too specialized, even if the different groups are interdependent.

It is this process that gives the strange, almost contradictory impression of both bushiness and an evolution that seems to be on course which seems oriented. If the selection pressure is the same – the success of technicization – it appears in the waves of disappearance during climate change, or in permanent nomadism, because technicization increasingly closes the technical group in on itself, allowing it to be nomadic. This is the process of "insulation" according to Sloterdijk.

What is happening in these increasingly technical groups? In *The Lion*, Kessel describes the life of a Maasai (Kessel 1959). To become an adult, a young man must fight a lion and kill it. This is the establishment of a terrible selection system. It is understandable that a technical mastery as impressive as that which allows a young adult primate to fight a lion victoriously can allow a ruthless selection and give such a group a form of invincibility compared to any other animal species (except its own). This is an ancient Maasai specificity; in other tribes, the passage to adulthood requires other initiation rites. Here we describe a general selective system, which is expressed in various ways, in the framework of modern humanity, after the long evolutionary history of the hominin lineage.

We do not see why, as soon as the combination that combines bodily technique and equipped technique with the social organization that favors it is put in place, there is not a process that can be described as autocatalytic, by analogy with chemistry, which, from generation to generation, is constantly pursued, sometimes accentuated, sometimes diminished, or at least incessantly, on the scale of tens of thousands of groups in competition-cooperation. Within the groups themselves, the difference may seem minor between those who master the techniques better than the others, while environmental hazards may sometimes exceed the technical variations between the groups. In the long run, a better technical mastery allows a better adaptation to climatic variations.

15.6. The relationship between technical behavior and biological evolution

Whatever happens, biological evolution cannot go beyond the rhythm of reproduction, the appearance of mutations nor genetic recombinations, which is a partly random process. The steps of change are not very compressible, although variable. This means that a biological/morphological change is not perceptible to the populations themselves; the same is true in modern times, where a kind of anxiety of "racial degeneration" can be detected as a result of medicine and a less violent education, whereas it is not detectable on the scale of a few generations. This kind of concern can be found as far back as we can go. The degeneration of youth is recurrent, in all civilizations, as soon as it was possible to write it down (Babylonians, Chinese, Greeks, Romans, etc.). This clearly shows that human-technical co-evolution is not understood: we know and perceive what we lose; we rarely understand what we gain.

There are at least three hierarchical levels of biological selection. In a group, an individual or couple, if they have sufficient autonomy within the group, could, with their technical abilities, be more successful and have more children who become adults. At the group level, better coordination or stimulation or emulation, if it makes the group more successful, may lead it to "swarm". At the level of a territorial space, exchanges between groups, whether individuals, males or females, or techniques, favor "gene circulation", mixes, innovations. At the species level, the more a limited territorialization effect is established, the greater the risk of reaching an impasse, but "disruptive innovation" remains possible. The relationship between each level of scale is not simple either, and geneticists have already modelled some cases. With the advent of the technique, which is transmitted more or less independently of genes, a new selection logic is added, which does not simplify the analysis.

There cannot be a causal relationship between technical behavior and biological evolution, as it is a probabilistic relationship, in the succession of generations. As we have seen, technical mastery leads to changes in living conditions, and thus to changes in selection conditions. This does not mean that any form of adaptation is imaginable, nor does it mean that it is possible to predict what can mutate and what will be selected.

The evolution of terrestrial mammals that became cetaceans comes from a return to marine life by quadruped animals. It has even been possible to follow the sequence that took place in the region now represented by southern Pakistan. There has been an extraordinary adaptation to swimming to the point that, in ancient times, a dolphin could be mistaken for a fish. Yet cetaceans have not adapted to breathing in water; they breathe aerially, even though some are capable of diving to surprising depths, which implies a technical mastery of the body's breathing. Likewise, they have not changed the way they reproduce, and the young suckles its mother. This can cause real problems; the dolphin's infant, not being able to control its breathing, can drown at any time, which requires incessant maternal vigilance, which has been selected. The other selection pressure is swimming, and it is easier through successive mutations to have webbed legs than to have lungs that turn into gills. Similarly, the tail fin of cetaceans meets the hydrodynamic conditions; it is not vertical like that of fish for reasons of initial morphology. The difference between a human and a dolphin is that they come from two lineages that have pursued different *choices*, verticality and technical differentiation of the hands versus horizontality and happiness in the water.

As for the problem of the babies, the mothers were so caring that they evolved to be able to stay awake for many months; and the huge size of their brains, selected to address it, allowed them to "half sleep" while taking care of the little one. What is selected is not very predictable in advance, one would need to know the variability of each factor, their impact on other aspects, structural constraints, etc.

Selection applies to what is variable and related to survival and reproduction; it does not choose a theoretical "fittest". This makes it difficult to find the starting point for evolution. For humans, the evolutionary process that led to their development was almost impossible to imagine a hundred years ago. We see it today, where we still find the idea that it is the completed human being who invented society and tools according to a recent process of self-domestication. The tools are multifaceted and are transforming living conditions to such an extent that this will have selective effects. It has to be analyzed on a case by case basis, and the effects over time are difficult to conceive. On the other hand, it is conceivable that the current human neotenizations are responses to this socio-technical pressure.

For example, living in a very sunny region requires protection from the sun, by brushing the body and hair, clothing that offers protection through the layer of air between the body and the clothing, and protection for bare feet on overheated floors. In some cases visors may be necessary to protect the eyes. The habitats are more or less closed and expansive, in order to create shade and thermal inertia; these are shelters that concern the group. Protection from the cold requires similar strategies with one difference: orientation is less important. Protection against rain and wind can be asymmetrical since it is a matter of protection from the prevailing winds. All these protection techniques have the effect, in the long run, of selecting humans less on their resistance to the effects of extreme variations in temperature, humidity, light, than on their ability to make such protections.

All these protections seem to have appeared rather late, probably with the genus *Homo*, around the time of *Homo habilis*. We'll have to "acclimatize to" the opposite of what has prevailed until now. Abilities that we think are recent may be old. It is unlikely that the so-called human revolution before 50,000 BP corresponds to the invention of speech, which is already ancient. Whether there was something special, no doubt, it would probably be an improvement of language in relation to a new social organization, even if it remains difficult to think about. Another hypothesis is the establishment of a new process that would only reach its full extent with the end of the Paleolithic, which corresponds to the beginning of the Holocene.

For Steven Pinker (2015), who defends a Darwinian approach, this "instinct" of language was thus constructed and its practice (and selection) would have ended up inscribing this capacity in the body, in the same way as "walking on two legs". It goes without saying that its application *stricto sensu*, in consonance with our thesis, means that it was learned well, step by step, over a long period of time and that its use had a selecting function. It is conceivable that a form of operative, very simple language may have appeared, which mobilized the brain, and that by co-evolution between language and the brain, all its functionalities emerged.

If we conceive of language as a learned technique, and whose learning, transmission and use have ended up playing a selective role in its very use, we must try to conceive a succession of pre-languages and then protolanguages with approximate, vague, not very recurrent meanings, otherwise constructed at the beginning, because they are not based on inner representations, but on practices. Language is selected on the basis of its effectiveness in terms of the continuation of life. In its continuous improvement, it would integrate different modules of brain functioning, until the brain is organized to learn how to control speech. The process may resemble that of verticalization. In this case, the verticality of the *Australopithecus* could have led him to gibber and further technical experience would have played its part for both walking and speech. We would arrive at Jean-François Dortier's fourth hypothesis. What is shared is the bodily technique, associated with the equipped technique, which leads to the mental technique. It is therefore advisable to think of language, in its earliest beginnings, not as a means of thinking about the world, but as a tool for better socio-technical action. This tool could take different forms and by successive exaptation take on all the functions it has today, without being biologically engraved, but remaining highly selective.

The opposition that has taken place in biology between D'Arcy Thompson's "morphological structuralism" and the Darwinian approach speaks for itself. Darwin had grasped the difficulty of the question: how could a complex structure, stabilized after millions of years by specific functions, have appeared? Here again, for all techniques, we must practice thinking in terms of co-evolution, knowing that for language, it is a triangular co-evolution: a bodily technique, a mental/neurological technique, a cultural/relational technique, itself in co-evolution with the equipped technique. Does the first selection of the ability to articulate sounds that carry meaning come from a reason that is still difficult to imagine: seduction? Transmission of the use of tools? Expression of emotions? Social relations, politics? The "modern walk", and the ability to run, has required more than three million years of evolution, and consequently a "profound biological inscription"; the cultural, learned aspects of walking and running, however, are still evident. What about language? Why is it so different? Whatever it is, it is more recent.

There's an obvious difference. The first two technical fields are know-how in the sense of acting on matter, but in doing so, they transform the body, which step by step is selected for the ever-improved use of techniques and for the ability to improve the effectiveness of the tools. This is the reason for this co-evolution: selection based on learning abilities and selection based on the capacity to improve the tools. We could imagine a selection only based on learning. This would have produced a very adaptable monkey, without language. This is not what happened, namely because this transformation of the body is also a neurological transformation. As we have shown, *Homo habilis* and *Homo erectus* are beings in a state of metastability, although less so than *Homo sapiens*. Capable of learning to learn, they accelerate selection based on brain exaptation capacities. Mental technicality supposes an individuation at the mental level and the emergence of a different functionality. Through language, it is possible to talk about the world without acting on it. Nevertheless, talking about the world to others who listen to it modifies their representation of the world and their actions.

Terrence Deacon (1997), a neuro-anthropologist and researcher of the evolutionary processes of human cognition, proposed an evolutionary model that has analogies to the one we are seeking. Deacon proposes a relationship between language and the brain, which, metaphorically speaking, corresponds to that of a parasite and its host. In Buddhism, the conception of thought that comes and goes internally is considered a parasite of the brain for he who does not master it. From *Homo habilis* (or even before?) to *Homo sapiens*, language and the brain would have co-evolved, that is to say, in our approach, each would be selecting from, and selected by, the other. If we conceive of language as a technique which is invented, learned and transmitted, can we conceive of an evolution similar to that of walking, stone knapping, hunting, and the use of fire? In all these cases, the brain has evolved along with it. What language would have, compared to the "modern" language, the relationship between the broken stones of the pebble culture and the cut stones of the Upper Paleolithic? What minimal symbolic capacities, beyond the needs of a chimpanzee, could be useful enough to become a medium for selection? It seems consistent, in agreement with Terrence Deacon, that while these early pre-language forms became useful technical extensions of life, they exerted selective pressure on brain development.

In any case, if a protolanguage could have appeared, and developed, it must have had a use, even before the emergence of a "modern" language. The whole question is: "which?" As this question evolves over time, with the evolution of humans, it is a subject not only of analysis, but also of foresight. As with all complex functions, the appearance of techniques has taken place gradually. It has been shown that learning by students of the Acheulean method of stone-knapping was greatly improved when accompanied orally. This does not mean that the language was fully developed in *Homo erectus,* or that tool making was a factor in the selection of a "proto" language, especially if the latter was anatomically possible. As we have seen, without breath control, language cannot be accurate and stable. On the other hand, as soon as this control starts to be used elsewhere, it can be mobilized and be a source of selection for the articulated language. But the requirement for articulated language can also lead to the selection of breath control.

One last remark: in language, as in all other forms of techniques, the gap between admirable mastery and the basic level is considerable. We all know how to walk, run and control our breathing, but our level in these three practices varies greatly. The same is true of language. Parental influence, and more generally the influence of the environment, is evident. Nevertheless, lifelong learning and work show, in all these cases, the potential for considerable individual progress. If there is selection, it is at all levels, from the individual to societies, which isn't yet suitable to model.

15.7. The selection of neoteny

The first neoteny would be the conservation in adulthood of the vertical position already acquired by the small primate, with the selection of an immature great ape foot. Maintaining it corresponds to neotenic-type developmental shifts, or heterochronies. At the end of the *Australopithecus* bush-like lineages, that is with *Homo habilis,* it is possible that brain growth was also the result of a neotenic-type heterochronic process, bearing in mind that brain growth is not in itself an argument. It may be interesting to ask the question of the state of the baby *habilis* at birth, even if the archaeological data seem to be limited.

Let's go for the most likely and simplest hypotheses. *Homo habilis* had a bigger brain, thus the baby's brain increased in the same proportion. We have here a strong clue: the hips have widened, since *Australopithecus*, as an adaptation to bipedalism. Under these conditions, a true neoteny does not seem necessary, but rather heterochrony allowing the growth of the brain; bipedal walking has created a constant pressure to adapt to walking (foot, hip, spine), and one of the consequences has been the possibility of birth of babies with bigger heads. The increase in adult height associated with an enlargement of the hips leads to a possibility of "regulating" prenatal ontogenesis, leading to babies with larger heads in absolute and relative terms. This is a physical problem of surface/volume ratio. By chance? No. Physical law.

In the previously cited article on the female pelvis of a *Homo erectus*, dated approximately 1.2 million years ago, the authors add that this large pelvis shows an evolution in response to the growth of the fetal brain and conclude that neither adaptation to tropical environments nor endurance running were primary selection factors determining the pelvic morphology of *Homo erectus* (Simpson *et al.* 2008). This is true, but it is not possible that an increase in fetal brain size led to pelvic enlargement. This is a reversal of cause and effect, according to a Lamarckian approach which is not stated. Without other preconditions leading to this enlargement, birth with a large brain cannot take place; it causes death in childbirth or miscarriage. As we have seen, improved bipedalism is sufficient.

The question is changing: what can create a selection that favors babies with bigger heads, knowing that, even if it is possible, it is riskier, especially if the baby becomes more and more dependent? The answer may lie in a tendency favoring tenderness towards the baby and a tendency favoring dependent and fragile babies. Indeed, it is a general rule in mammals that babies are born non-viable. They are all, to varying degrees, dependent on their mother, and, for species living in groups, dependent on the protection of their mother. In higher primates, this dependence is very marked, and during the two million years between *Homo habilis* and *Homo sapiens* this dependence increased, bearing in mind that inventions such as the baby carrier modified the selection effect by improving living conditions.

This increased dependency led to a second form of neoteny, where the baby was born as a fetus compared to its ancestors and continued to develop for longer postnatally; it was both larger and less viable. The baby could develop longer in an environment that has become sufficiently protective. This trend must have been in place long before the appearance of *Homo sapiens*, known as the "biologically modern human", from the very beginning of hominins. In fact, at birth, the human brain represents only 25% of its adult dimensions and the same ratio is found in chimpanzees, while it is around 70% in macaques and gibbons. As for the shape of the pelvis and its strait, it has hardly adapted to obstetrical functions, but long before, to bipedalism and support of the viscera, and its evolution began with the first bipedal species. In *Australopithecus*, the neonatal brain dimensions seem to be smaller than those of current chimpanzees; the first neoteny of the biped is that the baby becomes more immature from birth. Hence the early pressure to invent a baby carrier, which itself allows increased immaturity.

Despite the pelvic width of *Homo erectus*, human brain processes would have been a handicap if fetal brain growth had not been slowed down (second neoteny); the membranes between the bones of the skull were wide open, allowing true compression at birth (Rebato *et al.* 2003, pp. 305–310). As we can see, this second neoteny is only a neoteny because the overall volume of the brain is very high in humans from birth. It accounts for 10% of body weight at birth, compared to 5% for chimpanzees, while brain growth after birth is proportionately the same for both species; the human baby is born with twice as much brain and even increases its lead. This again leads to the conclusion that many characteristics of the human species were already latent a very long time ago.

In the human species, and also in earlier species, there is a softening in the face of vulnerability and fragility. Faced with a newborn baby, most of us retain our capacity for wonder. How can this little being, so slender, with such a small face, big blurry eyes and such an imposing and fragile skull, exist and become, at term, an adult? Birth most often generates a sense of responsibility in the parents. This feeling is fundamental to the survival of species; and it is this feeling that, over a long period of time, becomes the evolutionary engine towards cerebral neoteny; it is both selected and a source of selection. At similar brain volume, increased immaturity will, paradoxically, be selected by the increased protection of the baby achieved by the technical and organizational capacity of the group. It is the same process for the baby dolphin. The chimpanzee's baby acquires a beginning of autonomy around the age of two years (which is already enormous) whereas it takes at least twice as much for a human baby.

Independently of the research and reflections of anthropologists and philosophers, we can perceive the strength of the myth of having remained a small child while becoming an adult. It can be found in the cinema where extraterrestrials are systematically described with heads whose proportions are those of a child under four years old: small face, big eyes, domed skull. E.T. by Spielberg is a kind of exception, because the face takes up the whole facial part of the head, giving an impression of superhumanity. For the other parts of the body, it is more ambiguous; the fingers of the hands are often longer and thinner, and often the legs are longer.

In comic strips, many of them Japanese, some depict heroes whose head to body proportions are those of a child between six and eight years old and not of an adult. The face is smaller and, above all, less elongated, the eyes are wider apart and larger, the skull is bulkier and "protruding". Unsurprisingly, these comic strips have a connotation of anticipation or science fiction and the taste for technique is omnipresent. Take the case of Yoko Tsuno – whose name is misleading, as it is a French-language comic strip – the heroine of a series of adventure and science fiction comics. It was created in 1970 by Roger Leloup, a French-speaking Belgian. Yoko Tsuno is Japanese and an electronic engineer, which at the time meant that she came from an ultra-modern world. The height of her head, which is rather round, is less than a sixth of the height of her body, while the average for *Homo sapiens* is around an eighth. In terms of body proportions, this young woman has a body less than six years old[3].

3 All the titles of the series have been translated and edited in English.

In the duration of his creations, Roger Leloup gradually changed the proportions, and while the character's head is still disproportionate, it is closer to that of the "normal" humans in the titles at the end of the series. This could correspond to the fact that Yoko Tsuno has become more classical, less "hyper-modern". Created by Peyo, the Smurfs, who are not humans, have a baby's body and a huge head, and everything is done to keep them something deeply childish.

More spectacular is the success of the Barbie doll whose magnitude provides us with both the projection of a parent and child on the expected future of a baby. This doll is a combination of three characteristics: a very marked neoteny, a description of an almost impossible being (feet too small, ankles too thin, neck too slender in relation to the head, a size so fine that it suggests digestive organs still diminished), and a sexually marked adult look (relatively large pelvis and breasts).

The success of this doll says something significant. It can be hypothesized that its success comes from its fit with an increased desire for humanity, even if it is a little misguided by compliance with gender norms (Debouzy 1996)[4]. Whatever the criteria, the Barbie doll is a picture of the desired difference between a human and a non-human great ape. Every effort is made to indicate an increased gap: the legs are longer and slimmer; the hips, shoulders and breasts remain relatively wide (symbols of verticality and fertility), the waist is narrow; as for the jaw or digestive system, they are limited; the head has the proportions of that of a seven-year-old child with an overall volume greater than that of an adult and a higher forehead; the eyes are larger, the mouth slimmer; the ratio of the height of the head to the height of the body is seven, which is much less than that of the average American woman of reference and this in spite of the vertiginous legs.

Despite its slender appearance, it has the vertical proportions of a youngster under 12 years of age. The message is clear: this is an adult that is increasingly removed from the original ravenous monkey by neoteny. The Barbie doll is to the average American woman of reference what the latter would be compared to a woman of the species *Homo erectus*. Mattel, the maker of this doll, proposes that little girls play with the doll which would be the adult of a supposed distant future, and this without any evolutionary

4 Marianne Debouzy was a teacher in American History at the University of Paris. The cited article is not translated in English, but the abstract.

theory explicitly announced. No classical painting, even as far back as Egyptian frescoes, overestimates the size of the head in relation to that of the body. The dream of being more than human by increasing the proportion of the head is therefore recent.

In fact, today, the more neotenic a child is, the more protected they are; the more protection makes a child neotenic, the more protection is increased in return. In the post-modern world, a 15-year-old, usually pubescent, is still very childish. Reading books from antiquity or the Middle Ages tells us that only a few centuries ago, a 15-year-old could already be recognized as an accomplished adult. Thus, in modern society, child welfare is being provided for a longer period of time than before; this is a recent development and yet it corresponds to a fundamental trend of the species. The simplest hypothesis: today's technology makes this possible on a massive scale. But is it sustainable?

It is likely that it has been millions of years since the crying baby has been better protected than the apathetic one or the one that becomes active as quickly as possible. The distress call has been selected because mothers are very sensitive to it. The same analysis can be done on the happy baby who also gets the graces of adults. Sad or happy, it has to be expressive. The neotenic baby can exist because it responds to a demand. Thus, a world that listens to everyone's complaints could in the long run produce ways of expressing demands that could take the appearance of complaints. The taste for neoteny has been important enough over the last two million years for a convergence to appear: birth in a fetal state of impotence apart from the ability to suckle; a head so big that it takes almost two years before the child can walk at about correctly; a baby who cries from the beginning of his life, capable of sketching various feelings; an infant capable of absorbing and understanding the human world in which he appears and in real interaction from the age of eight to nine months.

Thus, the second neotenization of hominids took place under a quadruple pressure:

– the pressure which is the result of technicity, itself gradually accentuated over time, and which enhances collective hunting and harvesting capacities;

– the pressure which comes from technique and favors increasingly efficient tool manufacturers;

– the technical and cooperative pressure, which creates an increasingly protective environment for mothers cutest and progeny;

– the pressure of defending the cutest and most touching which are often the most demanding children.

Since *Homo habilis*, or even long before, human groups were not made up of frightened beings constantly struggling against a hostile world and having to hide. They were coordinated groups, protecting their offspring and capable of hunting antelopes, deer, cattle, and other game while knowing how to defend themselves from lions, cheetahs, and other predators. A group of *Homo erectus* or equivalent hominins, 1.5 million years ago, became so active, so coordinated and so skillful that they could defy any other animal group. The danger lay in their hominin competitors.

It seems doubtful that the selection of neoteny stemmed from a kind of eugenics based on the *ex ante* selection of the strongest, the most capable, the most intelligent; in fact, the opposite can be said. On the other hand, the fourfold selection described above leads a population in this direction, provided that it is understood that this remains indirect and does not concern the individual alone. It is the group that is more able, more dynamic and stronger. We can see *ex post* that the knapping of stones in the late Acheulean era mobilizes the brain, and that this complex motor coordination is the basis of what we call intelligence. It is highly likely that hunting, recognizing animal footprints, making traps and shelters, maintaining a fire, controlling the cooking of food, the choice of consumables, methods of protection against cold, sun or wind are all contributions to brain development, and correlatively to changes in musculoskeletal morphology and structure. In fact, any technique, whether bodily or instrumental, mobilizes parts of the brain that contribute to intelligence, in its polymorphic sense. Thus, it is through an apparently anti-Darwinian neotenization, made possible by the technique, that a Darwinian selection has been achieved at the higher hierarchical level.

There is a third form of neoteny, and it is conceivable that it took place after the appearance of *Homo sapiens*. This is a psychic and behavioral neoteny. Indeed, prior to 100,000 BP, it does not appear that *Homo sapiens* lived any differently from other humans at the time. Between 70,000 and 50,000 BP, during this "human revolution", a new type of behavior, new material productions and above all a much more varied range of products appeared (Stringer 2012). They depend on the place and adapt to it. This is probably the beginning of a differentiation between a simple place (*topos*) and an inhabited place (*chôra)*, to take up Heidegger's distinction, which came from Plato and was insistently taken up by Augustin Berque (2015). From this date, *Homo sapiens* showed an increased capacity to adapt technically to an environment and to modify it. Through technique, they can create their world. We will see this capacity fully deployed later, with the emergence of agriculture. It once again confirms the anteriority of acquired capacities compared to the time when they are fully deployed. It is the new capacities for generalized exaptation that make this possible.

These new technical productions correspond to the most recent exit from Africa – after *Homo erectus*, *antecessor, heidelbergensis*, and the first of *Homo sapiens* – the one that would be global and lead the different human groups to spread over the five continents in about 40,000 years. Crossing the northern regions of Asia to the east took more than 30,000 years, or about 2 km per year. This was not a raid but a slow and generational spread. There were breaks; the crossing led to America. Archeological evidence suggests that some passages required some cooling, then warming (crossing the Behring Strait, then Alaska to the more southern part of the American continent). This shows quite remarkable adaptive capacities, which must nevertheless be put into perspective, since changes were slow on a human scale. Even if the antiquity of the qualities demonstrated is most often underestimated, for the reasons already mentioned, it is highly probable that a psychic neoteny took place after the acquisition of the modern morphology.

One hypothesis is conceivable. It still comes to us from the analyses of Maurice Pradines, Frank Tinland, Peter Sloterdijk and François Sigaut. They admitted that language appeared in parallel to technical development by analyzing what the technique needs to be able to be transmitted. While it is indeed difficult to define a cause-and-effect relationship, the fact remains that language is what makes it possible to specify reasons and motives, to convince and to gather. You can copy know-how; the more complex and time-consuming it is the more time it takes, if you do not understand the reasons and motives. The four Aristotelian causes that lead to a result are indeed those needed to transmit a technical process; what Aristotle analyzed is what language says.

This characteristic of language, which makes it possible to speak about a technique and to transmit it, is the one that the chimpanzee does not reach. Experiments conducted under the direction of Dietrich Stout (2015, 2016), at Emory University in the United States, suggest that starting from the Acheulean techniques, transmission through language accelerates learning. More specifically, Anne Smyrl (2014), at the University of Colorado Boulder, shows that, for post-Acheulean techniques, learning by observation and imitation was no longer sufficient to transmit technical know-how. Verbal learning had to begin before the post Acheulean era, specific to *Homo sapiens,* and it is this technical practice that selected a new behavior over time.

These studies corroborate Maurice Pradines' analysis. To maintain techniques more complex than those of Acheulean (typically the so-called Levallois technique), language became necessary; for the Acheulean techniques themselves, the language was already a real asset. The technique may have been a selective factor in oral development. The appearance of a language by which the individual is able to explain reasons and motives – the Aristotelian causes – and to describe objects, their parts, their functions, produces a new selective medium. It is a symbolic medium that can be conceived as a causal agent of an acceleration of technical development. According to these studies by American anthropologists, language would have been established in its modern form anyway before the Mousterian and during or shortly before Acheulean. This does not mean that there was no language before. The evolution must have been progressive; "modern language" includes conditionals, purposes, reasons, the search for causes, a diversity of links, relationships, positioning in space, temporality, possession, desire, feelings, etc. and it is a language that has been developed in a progressive way.

15.8. Towards the human being: convergences and co-evolutions

We understand that the biological evolution that led to humans took place within the framework of a co-evolution associating bodily technique, equipped technique and biology. Language, primarily a non-equipped bodily technique, would necessarily lead to the mental technique. It was gradually necessary from a certain development of the equipped technique, because without it the technique could not have gone beyond the Acheulean stage;

nevertheless, it is doubtful that it was technical use that was the final selection condition of modern language. Dessalles (2007) suggests that the final selection pressure was the need to create strategic alliances in growing groups; this is a selection by ricochet, for the growth of groups is itself a consequence of technical progress and the biological neotenic response with growth of the cerebral cortex. If strategic alliances become necessary for the survival of a large group, language selection can take place and in turn it will amplify the possibilities of technical evolution. There are therefore many moments of "crisis" when convergences are selected.

Since the major program on the human genome, the search for the genetic functionalities of all DNA sequences has become possible (Kellis *et al.* 2014). In this context, evo-devo studies have shown that it is possible to increase or decrease the expression of certain genes just by modifying regulatory sequences. For example, the switch to bipedalism is associated with the loss of an ancestral activator that controls the expression of the hind limbs, but not the forelimbs. Gains and losses of regulatory elements can alter body patterns. Finally, evo-devo studies agree with d'Arcy Thompson and Darwin. The evolution of the forms can be obtained by modifications which are tantamount to switching regulators on or off, which thus accelerate or delay local developments by transforming the forms according to the rules of physics and mathematics.

It is known that the human genome has lost several hundred DNA sequences that play a key role during embryonic development and amplify, switch on or switch off genes, compared to the genome of nearby great apes, but also to mammals in general. Even the growth in the number of neurons in the brain seems to have been amplified by the disappearance, during the third phase of neotenization, of a switch during ontogenesis (Indjeian *et al.* 2016).

Changes in gene expression regulation may be likely to produce phenotypic effects while preserving the viability of the organism. They are known to cause interesting evolutionary differences in other species. However, there are molecular events in human DNA that can produce significant regulatory alterations, such as the complete deletion of sequences otherwise conserved in chimpanzees and other mammals. To date, 510 such deletions have been found in humans, occurring almost exclusively in non-coding regions in the vicinity of genes involved in the signaling and neural function of steroid hormones. Paradoxically, in order to achieve a more developed brain and nervous system, regulatory genes that limit this development must above all be suppressed. It is indeed lethal under natural conditions (McLean *et al.* 2011). A gifted and overdue baby chimpanzee is of no interest to the community ... but in the human and pre-human environment, such qualities were selected as long as they met the requirements of technical use and were possible thanks to the initial technical protection. A reciprocal feedback allowed this body/technical nervous system co-evolution.

Evo-devo research has also shown parallel processes in the different lineages under similar selection pressures (Reno 2014). Historically, in biology, a homology was separated from analogy. The wings of birds, bats or pterosaurs, all tetrapod vertebrates, have similar functions, in that they function for flight, but are not homologous, that is they do not originate from the same genetic and structural structures. The bird's wing is carried by the entire forelimb (the arm in humans), the bat's wing is carried by four fingers, and that of the pterosaurs by a single finger. We find the same problem when comparing the fins of classic fish (osteichthyes), sharks (chondrichthyans) or cetaceans.

All these phenomena of convergence, parallel evolution and even reversion are now grouped under the term homoplasy. By confronting the history of evolution on many branches with "hominoid radiation", thanks to the refinement of the comparative anatomy of the apes, we come to the conclusion that homoplasy played a very important role in the evolution that gave rise to humans (Reno 2014). *Ardipithecus ramidus,* in this approach, brings important elements. It is quite neotenic in the sense that the adult is close to a small bonobo standing vertically, both suspended and erect (Lovejoy 2009).

The concept of neoteny is close to that of the loss of regulation; it is indeed to speak of the withdrawal of a differentiation to give way to a possible non-differentiation, which is differentiated in and by technical use. Technicality opens up numerous differentiations and in response the biological aspect adapts by playing on non-differentiation and the capacity to use differentiated tools. The slightest de-differentiation is associated with the integration of the use of tools in potential and conjunctural differentiation. The body and the mind are always undifferentiated as a response to a technical solution, and it is the potential capacities that are amplified with the technique. Differentiation becomes "gestaltic" because technical use, and anything that promotes learning, can be selected. This has been one of the secrets of evolution towards human beings for millions of years. Many of the selected mutations are in fact neotenic by reverting to non-differentiation.

In the present state of our knowledge, it does not yet seem possible to decide precisely when our ancestors acquired speech and the faculty of language in its doubly articulated representative language sense. On the other hand, we know that it cannot appear before what makes it possible. Michel Foucault (1994) has shown that the same process is visible in the modern evolution of language. It requires, at the same time, the production of stable and calibrated sounds, the coordination of the laryngo-vocal duct associated with the control of speech articulation (teeth, lips, tongue, palate) and their association with the mastery of syntax and semantics. In addition, recursivity of brain function must be implemented[5]. Intellectual activities provoke the development of neuronal connections and the neuronal networks that have become more complex allow a richer intellectual activity (Hauser *et al.* 2002).

While we cannot know exactly how and when language appeared, we have seen that it is possible to understand how and in what order the different functions that allow it were selected, and approximately when it acquired all the qualities of a complete language. In the same year, two authors set out in different ways to describe how the human species emerged (Dortier 2004; Gärdenfors 2003). Both promote imagination and the production of representations as what defines the human being and would become the "true property of man". Peter Gärdenfors analyzes the order in which the different characteristics of thought and consciousness should logically appear, systematically seeking the order of the necessary conditions. Numerous publications now point in this direction.

This order is consistent with the role of technique (tangible and production resources/tools) as the first factor. For example, to have sensations is a necessary, but not sufficient, condition for perception; to perceive is a necessary, but not sufficient, condition for being aware of objects; to be aware of surrounding objects is a necessary, but not sufficient, condition for being aware of oneself. Being self-conscious is a necessary condition to speak a doubly articulated and representative language, but

5 Recursivity is what makes it possible to refer to the very object which is the subject. This is the case with any dictionary, any encyclopedia, and language.

contrary to what many philosophers still believe, it is not sufficient; self-consciousness comes long before modern language.

Seven years later, Yuval Noah Harari (2015) re-emphasized imagination as the foundation of humankind and what will distinguish human communities from all other forms of animal communities. In fact, these three authors develop in a phylogenetic and historical approach what Gilbert Simondon (2008) had well identified in his 1965–1966 course entitled "Imagination and invention". Imagination, which the rationalist thinking of the last four centuries had relegated to the position of "a chattering of the mind" today becomes the "substance of the psyche".

For it is not only by his brain – or the functions it carries – that the human being is distinguished, or by his body configuration, or his verticality, or his ability to learn, sing, paint or speak, etc. It has taken more than five million years, a few hundred thousand generations, to go from a superior social standing ape – having already, to a reduced degree, "theory of the mind", consciousness, self-awareness and empathy – to the human being we know. At the stage of our reflection, this seems coherent, given the number of stages, of which at least three are biological neotenization. It seems illusory to believe that it will be possible to define the human from a single functionality.

15.9. *Homo sapiens*, a convergence of multiple capacities

The transformation, even if it is based on qualities that are all present like "signs" in the chimpanzee or the bonobo – as well as in the orangutan and the gorilla – and therefore in the common ancestor, shows a convergence of developments that all seem to be driven by the appearance of what we have called body and tool technique. They create new conditions of existence whose resulting selection pressure leads in the same direction, toward greater bodily technique and greater tool technique.

As a technique is something that can be learned and takes time, it involves the collective. The implementation of a technical approach is done in a collective interaction, and increasingly so as it becomes more complex. Technical development takes place with a development of collective exchanges and consequently with a development of socialization processes; the intensity of the exchanges, already important for chimpanzees and bonobos, is increasing. More techniques lead to more learning, more coordination and more exchanges between individuals in the group, even if the technique *in fine* depends on individual mastery. Hunting, scraping and cutting techniques cannot be strictly individual.

Thus, between four and six million years ago, a selective global process was set up to promote technical development (physical and technological), specific learning, in interaction and exchange, which required communication tools. After the "chopper", the "Lomekwian culture", then Oldowayan, then Acheulean, it is possible that this succession was only a small, simplistic overview of real attempts, those that failed and those that succeeded. We still have few traces.

It seems very likely that the first capacities for representation emerged at this time. Gilbert Simondon, during the 1960s, at a time when imagination was still considered a quasi-parasitic function, undertook fundamental philosophical research based on the international scientific and philosophical knowledge of the time. In 1964–1965, he gave a course on perception in which he addressed the relationship between perception and imagination. It was at that time that he established the concept of "the intra-perceptive image in the perception of forms" and constructed an overall perspective that allowed the role of imagination to be posited in a new way. He develops an integrative approach, which could be called the "Simondonian method", bringing together diverse knowledge from different disciplines in order to understand the "genesis cycles".

This course on perception is also a course on the history of theories of perception and the history of philosophy from the perspective of perception; it shows the evolution of the status of perception in philosophy and theorizes the role of perception in philosophical choices during the history of philosophy. This text leads to an interactive understanding between perception, action and imagination. For example, Jean-Marc Lévy-Leblond (1996, p. 146) observes that relativity was used to geometrize gravitation. However, it is by degeometrizing the latter that particle physics has brought it closer to the fundamental interactions of matter. How can we understand this double, contradictory tendency, already present in philosophy at its beginnings? Global thinkers geometrize, microscopic thinkers think about interactions. At the macroscopic scale, geometry prevails, while at the microscopic scale, one must think about relationships; this is already the case with the presocratics. This is not specific to modern physics, but to our perception. If we hadn't had the perception that we have – "us" being understood as the humans who have been able to achieve this scientific revolution – how could this step have been taken? We therefore have long-ignored capacities, which are revealed in collectives, in the face of new problems that at times produce them.

The following year, Simondon gave a course on imagination and invention, taking the previous theses for granted. Two years later, he gave a course on invention in techniques. It is not the place here to analyze Simondon's entire body of work concerning the transition from animal to human in terms of the role of the perception-imagination-invention triptych, which should henceforth be thought of according to a genetic unit, each being a phase of the same genesis process. Let us simply note that image and imagination are presented as vital, fundamental functions from which the psyche can emerge, which Simondon does not limit to the human being alone. "What characterizes the image is that it is a local, endogenous activity, but this activity exists as much in the presence of the object (in perception) as before the experience, as an anticipation, or afterwards, as a symbol-souvenir."

Simondon shows (as Goethe has already asserted) that "motricity precedes sensoriality, as a long-term anticipation of conduct"; "spontaneous behaviors are a permanent and necessary anticipation of perceptions"; "motor equipment precludes sensory equipment"; "the organism is a set of behavioral patterns as clearly definable and taxonomically valuable as the shape of the dander, the number of claws, etc."; "therefore there was a real biological basis in the imagination, prior to the experience of the object"; "the psychic level of activity refers to an environment that has already been explored and organized biologically, i.e. to a territory".

The basis for an image production capacity existed in the ancestors of the bipedal primates; it could not be mobilized as a selectable capacity for the simple reason that the protected territorial space, the environment, was not yet sufficient. Chimpanzees and bonobos have already moved beyond the primary stage of immediate response to needs. The technique will offer the possibilities of an enlargement, moving to the development of an increasingly broad psychic mode and then to the formal mode that establishes relationships, analyzes the reasons and understands the motives.

It is a continuous development, from the most primitive pebble culture to the Lomekwien, the Oldowayan, then the Acheulean, then the middle stone age (which would correspond to the Mousterian). The first two would be carried by the *Australopithecus*; the third by *Homo habilis*; the Acheulean from *Homo habilis* to *Homo erectus* and *Homo heidelbergensis*; the middle stone age to *Homo neanderthalensis* and *Homo sapiens*. Everything happens as if after each identified technical novelty, and after a certain time, a biological evolution is visible. Biology responds with a delay, because of the inertia of biological evolution; and it is not because biological evolution has taken place that technical regressions are impossible. The imaginary function increasingly develops to the point of playing a dominant role in invention, when language is capable of transmitting the imaginary itself.

The evolution of communication towards language must have followed the same pattern sequence; since language is a bodily technique, that is both biologically based and learned in a psycho-socio-technical context, it is this context that has been the driving force. We are "artificial apes" in the words of Timothy Taylor. It is very likely that language, in the modern human sense, has reached its present functioning with this human revolution between 100,000 and 50,000 BP, at least one million years after language, in a broad and fuzzy sense, was a factor selected by technical development (Acheulean, then middle stone age/Mousterian period). From selected, it quickly becomes selecting[6].

Around 100,000 years ago, humans had long known how to use ochre dyes to paint their bodies. This indicated a self-awareness that no great ape is capable of. Yet, for about four thousand generations, *Homo sapiens* have lived, apparently, like other hominins. What could have happened, if not a final evolution of language, based on the amplification of recursive cerebral functioning? This recursivity finalized the evolution of language, making it doubly articulated, representative, to be learned, and subject to incessant contingent evolutions. The human brain has acquired a plasticity that will enable it to conceive and learn any natural language, artificial languages (such as those of techniques, science or law) that even allow it to transcend space and time, highly formalized technical approaches, and complex social constructions.

6 In that time from being selected to becoming selecting it could take thousands of centuries Think about the modern world … now less than thirty years old, and even less and less.

When language reached its full capacity for recursivity, coordination and control of speech articulation and their association with syntactic and semantic mastery, it is now acquired through the teaching by parents and the group; a culture of generalized transmission has stabilized. It is through cultural capitalization over a good number of generations that language was constructed. Around 200,000 years ago, humans spoke, even if their languages were incomplete from our current point of view; they still had to evolve as "complete" languages.

This evolution makes it possible to understand the dialectic between "thinking in images" and "thinking in words". The difference between chimpanzees and humans is not only in one of the two fields of what thinking is; it is in both, which have developed step by step in interaction. Today we are witnessing a return to the recognition of the importance of technique (that is to say, of action) and of the image (that is to say, of the imaginary, in thought) in the knowledge that twenty-six centuries of philosophical research had given priority to what could be said: language. This language-imagination co-evolution is both extraordinary and understandable. All primates, including humans, have sight as their dominant sense. Humans developed hearing and speech throughout the process of step-by-step amplification of the imagination, putting them in very close interaction. The so-called doubly articulated language is a representative language; it is based on sound, which is intertwined with sight, associated with all the other senses.

In this context, neogenesis and neotenization are shown to have been the biological means of response to a selection pressure constantly pointing in the same direction. Neotenization and exaptation – which in evo-devo language is described by "homoplasy" – have become human characteristics, which exist in and through the interaction between the biological and the sociotechnical, the biological having found the necessary recursivity through exceptional brain development. It cannot be said that sociotechnics and culture have simply been added to biology; the complete human appears when the neurobiological becomes plastic, capable of absorbing and responding to the demands of sociotechnics and culture through learning.

Brain modularity shows that mental capacities were set up sequentially during phylogenesis by increasing the neocortex; all the brain modules are in strong interaction. The human being is endowed with multiple, interacting and highly differentiated qualities, according to a complex probabilistic brain ontogeny based on genetics and epigenetics. What emerges is more complex than simple genetic causation. The human being remains phenomenologically multiple, since individuation is never complete. The ability to learn, including language, though unique, is thus context-specific and subject to contingency. Each individual learns what his or her context gives, and if the context impoverishes, it impoverishes too. Individual development depends contingently on the context in which the individual is born.

15.10. The ultimate technical step towards the human: mental technique

We have seen that it is from the bodily technique that a process of positive feedback has been set in motion which is constantly developing bodily technique and tool technique. Technical development creates new conditions of existence, which are the basis for the selection of any mutation or biological variability suitable for the maximization of technique as a means of improving individual or group competitiveness, or even on a species scale.

The bodily technique is constantly being refined, by internalizing techniques, accentuating the efficiency of the tool technique. From a certain level of complexity that can be defined at the time of the Acheulean culture, learning requires more than emulation and copying. It requires a communication that can at the beginning be "signed" or "iconic", that is to say a form of "primitive" language. Language is still simple, but it can still convey logic and describe gestures. At the beginning, the aim is to draw attention to the most significant points, objects or parts of objects relevant to the situation. A very strong selection appears based on a potential to communicate beyond that: plans, reasons, motives. This leads to a propositional language. It turns out, in fact, that an evolution of the brain has been possible to design what Peter Gärdenfors calls "detached representations" or what Gilbert Simondon calls "images".

The representations are detached in the sense that they acquire autonomy and can recombine, which is consonant with the Simondonian analysis of the image cycle. The evolution of the brain, during the seven million years of evolution of higher primates leading to humans, has led to the possibility of producing such representations. Neurologists are making progress in deciphering such productions. Will we ever know "how the brain does it"? It is difficult to deny that it is indeed this capacity for separation and combination that will allow for the establishment of an articulated thought that will then be able to find language as a support according to a selective process that we are attempting to analyze here.

The language that emerges, in stages, is therefore certainly a bodily technique, but it reveals a new technical field: mental technique. Representations of time and space, of events, of the conditional and the possible, of motives and causes, of links and relationships, will be able to be codified and transmitted. The mental technique, which is to associate representations, symbols, signs, codes, in a virtual inner reconstruction of space and time, will find an unexpected medium in language.

The mental technique was probably built autonomously, without any bodily technique, even if we have difficulties in conceiving it, because it had the unexpected medium of (pre and proto) language at an early stage. Corvids, for example, can count, at least up to four, without the medium of language. Wolves, chimpanzees and bonobos are able to plan an action without language. On the other hand, it is difficult to imagine a mental technique that would first require a tool, because the tool for writing language (writing) seems to have been around for a long time afterwards; this ability to write has not been able to play a major selection role in the very short time (according to biology) that separates this invention from our time.

Counting, imagining and anticipating are a mental technique, but it appears that one cannot go very far without language, which is a bodily technique at the service of a mental technique. This bodily technique has created the selection conditions for a biological evolution allowing for a fluid and complex language. We can deduce from this that the beginning of language, in its most archaic sense, a pre-language, is really ancient. Just as an outstanding hunter species, thanks to the technique, only became champions in all categories after a few million years of hunting, so a speaking species only acquired a general capacity adapted to the needs of transmission after hundreds of thousands of years of selection.

Humans who appeared between 100,000 and 60,000 BP discovered a new power. They had acquired the power of a gesture which was precise and guided by its effects on matter. By this power on the gesture, the bodily technique is a creation of a space of freedom. Through the mental technique, they discover the possibility of internally constructing a universe of relationships, which can be structured according to a virtual spatiotemporal logic in which action plans can be imagined and implemented. The power of the inner construction (the mental technique) on the body technique is efficient. To think, to build internally, to translate into words, to transmit the construction and to see, in fact, its effect on another who is learning, such is the first discovery of the accomplished human being. That this comes from the neuronal development made possible by the evolution of bodily techniques and tools seems obvious. On the other hand, to think that it can be deduced from these corporeal and equipped techniques is an error of principle. For the link cannot be a relationship of cause and effect. The link is probabilistic, because it implies the logic of the living (see above). A necessary condition suggests possibilities; but as long as they are not visible, no one can know what will appear. Thus, it is possible that there are features of language that have selected language by exaptation, such as the taste of Dunbar's "gossip", or the political sense accentuated, compared to chimpanzees, by the size of the groups (Dessalles 2000).

We have already seen that the bodily technique seems somewhat magical, in the sense that to describe it totally according to a logic of proposition is out of reach. This magic cannot be perceived by the being of the bodily technique, who experiences it directly. It is necessary to be in the mental technique to perceive, thanks to its recursivity, these two very different levels, which open each of the spaces of freedom. When the mental technique can blossom, it means that there is identity in the relations between, on the one hand, the thinking subject and his thought, and on the other hand, the acting subject and his action. The "magic" of the modality of functioning of the mental technique then fully appears.

Without knowing how thoughts, images, representations come, it is possible to use them as objects. We understand that the action of the body on tools works by material connection. However, we perceive directly, from the end of the tools, what should be surprising. On the other hand, to discover that we can only think and say it, and then provoke bodily changes on ourself and on another who returns the effect, is magic. Because it also works on the emotions directly; to say your emotion is to transmit it; it is the power of words, images, the "objects of art". The "manipulation" of others becomes possible through language, which is, from the point of view of the body's techniques and tools, magic. Even music can be codified, and the codification brings new creative possibilities. This would come to be measurable with the effect of music notation, which was introduced between the 9th and 15th centuries, very recent.

Access to mental technique reveals that bodily technique is also enigmatic. Alone, without mental technique, it cannot say what is magic in it. The access to the technicization of the mind, that is to say, to the use of mental tools, associates a non-conscious, non-mentalized functioning, with the possibility of transforming "mental objects" into tools of action, first mental, then corporal, then equipped. The human being is built on three levels of interacting technical action. The mental technique, when it becomes conscious, sheds new light on the other two, which can now reveal their meaning. No philosophy of technique can be conceived in the absence of mental technique. It is one of the consequences of mental technique to make belief and thought not only thinkable, but transmittable.

The mental technique proceeds from the discontinuous insofar as, like in the relation to technical objects (tools, instruments, utensils), it takes individuated mental objects, be they images or words; it articulates them by defining the links, the relations between each mental object; these links and articulations are mental objects; finally the user of these mental objects, if he considers himself, is also defined as a mental object. This self-enclosure of the process (recursivity) is in contradiction with discontinuity and the principle of the excluded third party. The consequence of this is that everything that also has to do with continuity is defined as a mental object. This interaction between the part and the whole is a deep source of indeterminacy, and consequently of anxiety. To get out of this anxiety, the construction of an understandable reality given as coming from elsewhere seems to be an obvious solution. Subjectively, it does come from elsewhere. It is a vital necessity, an ultimate referent, a "for what purpose" that puts everything back into dialogue, as two individuals would be.

I will give an anachronistic but significant example here. When the company Apple produced its first computers adapted to the needs of non-computer users, one of the features was that physical movements could be obtained from the computer from the "icon" effect. The person clicks on an icon representing the contents of a floppy disk, moves it into the trash can icon, and the disk is ejected. This simple operation seemed like magic to the first users. A mechanical, physical movement was obtained from the relation to an icon, that is to say a "virtual" being. Today, it seems obvious; at the time, it was a revolution, not for the computer scientists, but for the users.

Because it meant that the relationship with the tool could begin to resemble the one that takes place between two people, and of course the one that everyone has with himself. A machine could operate according to the logic of the mental technique!

We had forgotten that the verbal relationship between two people is "magic"; indeed, it connects meaning with the relationship of cause and effect. In fact, starting from thought, everything can be put in relation, in association, in entanglement. Lévy-Bruhl[7] noticed this clearly when he fell for it at the time. He thought that for "primitive people" everything was magic. Yes, everything was magic, but it was also technical. The magic of "primitive people" is the recognition of the magic of the effect of constructed thought and its transmission beyond a mere physical effect. A "primitive person", who discovers that the individual who calls himself civilized and wants to understand him is not aware of the real magic of the direct effect of the sense on the relations of cause and effect, will want to deceive the researcher and we understand it well.

Chimpanzees are able to thwart researchers' efforts to understand by not seeking to do the work that researchers ask them to do. All you have to do is please them to get the reward, since conveying meaning is more than just agreeing to get a reward. Similarly, many children seek first to please teachers rather than to understand what they are being taught. Most teachers fall for it ... at least for a while. It is believed that meaning has been passed on because the effect is correct. In fact, it is the concern to please the person who wants to transmit that wins out. By losing sight of the magic of the transmission, one can lose the need to verify that it is indeed the meaning that has been transmitted. We confuse meaning with the use of a toolbox.

To transmit meaning and to see it materialize in another is, in fact, profoundly magical, enigmatic and barely comprehensible apart from the call to transcendental or cosmic forces, which in any case are powerful and mysterious. This awareness of the magic of the effect of the conscious act on materiality, be it corporeal or equipped, is an inner earthquake, an

7 Lucien Lévy-Bruhl's ethnological works, in six volumes, are all accessible in the Gallica digital library: *Les fonctions mentales dans les sociétés inférieures* (1910), *La mentalité primitive* (1922), *L'âme primitive* (1927), *Le surnaturel et la nature dans la mentalité primitive* (1932), *La mythologie primitive* (1935), *L'expérience mystique et les symboles chez les primitifs* (1938). In English, we can find at least: *How Natives think; Primitive Mentality, The "Soul" of the Primitive.*

earthquake experienced by some children who are mainly focused on logic and metaphysics. This is what is commonly called "the mystery of existence". It appears, for those who have been able to exchange with people close to the so-called "primitive" thoughts, that this consciousness is very intense in them. During the millennia that will follow one another after the Upper Paleolithic, some peoples will develop an anesthesia of this consciousness of the magic of the human world. It will have a paradoxical effect, causing "the emergence of the civilizations".

To become aware of the possibilities of mental technique is to open up a new space; analogically speaking, looking at and recognizing oneself in a mirror is already approaching the universe of bodily technique, which is what a chimpanzee is capable of; to stand between two mirrors facing each other is to discover the possibilities of virtualization of reflection; it is to approach the shores of mental technique. If, among the multiple reproductions of the infinite self, reflected on each mirror we could move only one, that would be the equivalent of a mental technique. Everyone would be loudly proclaiming that it was magic!

We have domesticated magic, in the sense that we have become accustomed to it and we have learned from manipulating; we have therefore lost part of our self consciousness; in the 21st Century, we are in the process of rediscovering it in the technique that now achieves that magic that we thought was only in us. Some recognize that today's technology has now become magical. For decades now, science fiction – whether in books or films (*Dune*, *Star Wars*, *Matrix*, *Inception, Minority Report*, *Ready Player One* or the works of Philippe K. Dick, for example) – has often brought magic and "technologies" together. Here again, it was Apple with its iPhone that initiated this movement, avoiding the complexities of the technique and linking the iconic directly to the effect. The sensitive, meaningful relationship with the tool is realized; the network amplifies this impression; a new cultural mutation is underway (Jarrosson 2010)[8].

This shows once again that the beginnings of a technical mutation are built long before all the possibilities are present for its development. The

8 This French author is not translated in English. He said that magic was returning through smartphone and so on. But in 1965 two thinkers thought that magicians were returning; it is important to take is as a metaphoric vision. *The Morning of the Magicians: The Dawn of Magic (English edition)* from Jacques Bergier and Louis Pauwels.

emergence of speaking *Homo sapiens* capable of transferring meaning into an object, these are the beginnings of a possible revolution in principle, under the guise of other technical conditions, which first of all concern equipped techniques: the digital revolution. About 50,000 years prior, mental technique was already ready. It was first necessary to produce the necessary intermediate tools, which took time, many events and a lot of luck.

This is as far as we go here. But, in contrast to Simondon, we claim that magical thinking is the result of the mentalization of technique, as suggested by Jacques Ellul (1954)[9]. For the technique is, chronologically speaking, much older than magical thinking. The technique gains three facets: mental, corporal and equipped. Art becomes possible, and so does religion, in a more general sense than the monotheists. Linking what seems separate is human magic. From the projection of representations into matter, meaning can be inscribed. Writing, which is a technique equipped at the service of mental technique, inscribes meaning in the material and requires a reader. Drawing and painting, or even sculpture, began at least 30,000 years ago and are also techniques equipped at the service of mental technique. Any production of a meaningful object is at the service of the mind. Thus, most of the original tools also had the invisible effect of selecting the beginnings of mental techniques. Since a tool has at least one function, it implicitly produces a pattern and reasons; it stimulates and selects mental processes.

The mental technique has specific aspects that distinguish it from the bodily or equipped technique, even if it retains this identity of being a technique. The creation of mental tools seems at once difficult, elusive and immediate. Images, constructions can burst through; at the same time it is possible to seize and "domesticate" them. We will have to come back to this concept of domestication, as that it covers a deep reality of the technical universe that is only revealed through it. In mental technique, the whole of the conscious field, from which representations and units of thought are grasped, is the same as the one that is grasped. The emergence of mental technique took place in interaction with that of language. Speech is the interface technique between mind and body. It is, as we have seen, a *de facto* body technique; it is what has been built by necessity to discuss the reasons,

9 Simondonian thought is ontogenetic; it analyzes the genesis of the individual from the point of view of being and its phases. All this work is built on an onto-phylogenetic way of thinking, that is one that takes into account genealogical history: how did this happen in the succession of generations?

the links, the relations, the purposes; from there, it allows us to see a mental technique.

The mental technique existed before the equipped techniques, in an embryonic state, as we have just seen with reference to the corvids and the chimpanzees; the list could go on. It's hard to visualize it; it's guess work. Humans did not hesitate to attribute it to most beings, living or inanimate (animism) for the tens of millennia that followed. The use of material tools and the practice of language have considerably amplified, in the long run, the mental technique which acquires a real autonomy in humans with the double articulated language. The return to the material will be achieved through writing, but this is another story.

The advanced mentalization of technique changes the human being by installing it in meaning, by making him a builder and transmitter of meaning, the one who will become capable of inscribing it. This extraordinary mutation gives some indication of the excessive pretension that the 30,000 years following the last exit from Africa and the progressive expansion throughout Africa do not seem to have seen any major innovation. Jean-Marc Pétillon, at a seminar on April 30, 2007, is quoted as saying: "the equipment of the hunter during the Upper Paleolithic remained more or less the same for 30,000 years"[10]. Alain Testart (2012) insists in writing that this period would be marked by only three major innovations, at most, for 30,000 years, which seems limited compared to what happens afterwards.

Let's put these statements in context. It takes a few tens of millennia to absorb this burst of mental technicization; those who have experienced it may have been overcome with fear and enthusiasm, as are many children as they become aware of this ability. Hence the burial of the dead, the cave paintings, the magic, the search for contact with the world of meaning; the world of meaning, of combinatorial representations, of the imaginary, is also that of questioning. Finding a livable collective solution to this new capacity, source of endless adaptation and innovation, but also of lethal imbalances, could not be done quickly. It is conceivable that regulation came from social processes that mitigated excessive creativity through forms of coercion. This is the stage where *Homo* becomes "demens". The survival of the species stems from social regulation which has made it possible to channel this self-destructive "madness".

10 Quoted by Alain Testart.

As for the question of the relative technical stasis, let us not forget that the last exit from Africa took place during the last glacial episode which received different names depending on the place: Würm in the Alps, Vistulien in Northern Europe and Wisconsin in North America. *Homo sapiens* not only survived it, but came out of it stronger. It took advantage of the glacial episode to spread around the world and, at the start of the global warming of the Holocene, to embark on a profound conquest of the living world.

15.11. The technical inscription of the mind

Our narrative indicates that what we call the mind, that is to say, what a human being can act, build, invent, imagine, plan, confront his memories, mentally travel through time, tell stories, believe, hope, is inscribed in our technical capacities, whether they are equipped, physical or mental. We have demonstrated the extent to which they are intertwined, and how a phylogenetic analysis allows us to conceive their chronological development.

Francisco Varela (Varela *et al.* 2017) has helped us understand that there is not a disembodied spirit and that it is not just in the brain but in the body as a whole. Once we understand that the development of the mind is a consequence of a co-evolution between the individual as a whole and the tools he uses, it is possible to grasp that the inscription of the mind in the body is a consequence of this co-evolution. This makes it possible to understand why it is almost impossible to speak without moving, in particular the hands and arms, the head, the eyes, and the whole body. The body shows that the individual is thinking about the future or the past, and that he dives into himself for a conceptual effort. Many thoughts are associated with actions, and language itself, in its metaphorical structure, constantly refers to technical issues.

Can you imagine a baby who is not in contact with any technical problems? How will it develop? As Hélène Trocmé-Fabre (1987) put it, "I learn therefore I am". Such became the human at the end of the first cognitive revolution. It is from the learning of skills, transmitted by others who have learned them, that the human being has been built. While the brain and neurological activity have become so important as to give the impression of a form of autonomy, nothing in neuronal functioning can be totally separated from the body, which is the means of integration through action and perception. However, there is not just one way to learn; for the variability of human capacities, in the access to and use of techniques, seems infinite.

The inscription of the mind is corporeal and technical, which means that the mind is built in interaction with all the artefacts (instruments, tools, utensils, methods, procedures) to which it has access, which Lamartine had understood well. Yes, the artefacts stem from the human being that led to their constitution, so that humans can learn to use them. It will also be necessary to study how the inscription of the spirit is also social.

15.12. The construction of thought

In an evolutionary approach, we can detect different levels of thought corresponding to the process of individuation that leads to the human being. This supposes that we get rid of this overhanging conception that claims to see a major break between "animals" and the human (Descola 2013).

All animals have a mode of functioning based on the intertwining of action and senses, however archaic and limited they may be. Mammals are all endowed with the five senses that we are familiar with, with a predominance of smell and touch. Among mammals, monkeys have the most developed eyesight. The higher primates, the great apes, are probably of all mammals those which achieve a sort of balance between hearing, touch-tactile, sight, taste and smell, a balance such that there is a sort of fusion of the senses and an overall perceptive capacity. But some mammals are endowed with other senses such as echolocation (dolphin, bat), magnetoreception (humpback whale) or electroreception (platypus).

It is from this foundation that humans are constructed. The technique first allows an expansion of touch and some vision (body technique and equipped technique), then hearing, which will lead to language (mental technique). By positing that *Homo erectus* type species remain at the level of a protolanguage, its thought is a global perceptive thought. *Homo sapiens,* thanks to language, shift the balance towards a more auditory thought. His thinking is primarily auditory-touch in dominant component. The word, a sound entity, takes on a central status in the process of individuation, even if the visual inner representations dear to philosophers are present. The visual remains impregnated with the undulatory and inseparable aspect of the auditory and tactile. But the arts of drawing and painting develop first. The expression of intentionality remains “fuzzy” and global. It is much later that writing, especially alphabetic writing, will lead to a shift in this balance by first valuing the visual and linear causal thinking in those who know how to read and write. The brain of *Homo sapiens*, combining modularity and intermodular interaction, facilitates possible shifts in the importance of the different senses according to technical developments. It seems that Marshall McLuhan (2008, 2011) was the first to show the influence of technical innovations on the relative weight shifts of the different senses mobilized in ways of thinking. These shifts can lead to breaks in the “representations of the world”, behavior, modes of action and social organization. There are thus correspondences between certain technical innovations and the mode of existence, action and thought of groups of *Homo sapiens*, with inertia that leads to transformations in time specific to history. This can be examined in other narratives.

16

Emergence, Then Global Expansion

16.1. Rapid global development

Why is it that soon after the emergence of a new *Homo sapiens,* the human race was so quick to discover the world? About 10,000 years later, groups had already arrived in South Asia and Australia. About 20,000 years later, the first groups reached North America after crossing all of Siberia. Before the beginning of agriculture, all continents had been explored (Groucutt *et al.* 2015; Gittelman *et al.* 2016; Prat 2018; González-Fortes *et al.* 2019). As verticality led to nomadism, this was the "last" exit from Africa, which was also an expansion in Africa itself. Humans had dogs with them. These migrants spoke languages that became differentiated over time.

The speed of this expansion should be put into perspective. It was a particular movement since the fastest movement, the one that led to Australia, was estimated at about two kilometers per year. In one generation, the journey is just over fifty kilometers, that is a day's walk.

This suggests movements based on population growth and the random creation of new groups seeking a new hunting area. There was no conquest of the world. It was a search for empty territories, for food resources based on low population growth, as we have seen, of approximately a few percent per generation. It was a leisurely walk, not a marathon. Unless events were voluntarily preserved, after five or six generations, the memory of where the ancestors lived, more than 200 kilometers away, was lost. It could become mythical.

16.2. Great linguistic diversification

We know that 1,000 years after the separation of two groups, their language evolved in a differentiated manner to the point where they may no longer have understood each other, even though it is possible to learn the other's language. It is understandable that 40,000 years later groups may have had languages that seem to have been constructed differently. The fact remains that they were based on the same functioning of double articulation linking a phonetic structure to semantics. All human languages satisfy the need for exchanges that carry meaning in order to describe, search for and explain reasons and motives, establish links and relationships and even express emotions and the desire for the hereafter.

If it were possible to quantify and compare the speed of biological evolution and that of languages, we would find very large differences. One approach is to observe how quickly languages diverge. For example, by comparing Castilian, Catalan, Occitan, Provençal, French, Tuscan, Neapolitan and Romanian, it is clear, even without being an experienced linguist, that in a thousand years, languages can diverge to the point of being identified as different languages, implying a breakdown in communication and the need to learn the other's language. It can be said that, for complex living beings, the time required to separate into two species is roughly a million years. It can be estimated that the speeds of differentiation in linguistic matters are of the order of a thousand times faster than in biology. So, 65,000 years of linguistic evolution resembles 65 million years of biological evolution, equivalent to the multiple radiations after the so-called "Cretaceous-Tertiary" catastrophe, after the dinosaurs.

16.3. Co-evolution of cultures, languages and techniques

It is understandable that languages, cultures and techniques may have evolved greatly since that beginning, while retaining their basic structures and functions. That said, linguists have a specific difficulty: except for finding living fossilized languages, analogous to a Coelacanth or a Gingko, ancient languages do not fossilize, except through writing and by having kept a correspondence. Everything must be searched for from the viewpoint of living languages. Moreover, whatever the approach, evolutionary or not, it was important for linguistics to acquire its scientific autonomy, which has certainly happened since Michel Bréal (2018) and Ferdinand de Saussure

(2011), but largely after Darwin's publications, linguistics being a very young discipline[1]. However, when it was founded, it was tempted by the evolutionary approach, with "organic linguistics", even if it was probably too early (Blanckaert 2011). In particular, an evolutionary approach has been used for the study of Indo-European languages.

For nearly 40 years, even if the community of linguists as a whole does not believe in the possibility of reconstructing the entire evolutionary tree of languages, for the reasons mentioned above, some researchers have begun to work in this promising, albeit particularly difficult, direction (Joseph Greenberg, John Bengtson, Merrit Ruhlen, Derek Bickerton, etc.) since it is necessary to build databases that bring together the entire vocabulary of thousands of languages, some of which are in the process of disappearing. Throughout the history of science, researchers have always shown ingenuity and exceptional capacities for building up and collecting data. Why should linguistics be any different?

Technique and artistic activity have also evolved and have resulted in the magnificent works of parietal art. These works reveal an extraordinary closeness to us; we perceive that those who made them are like us. They have a sense of mystery, cosmic perception; they feel that they are of this world and yet beyond it; they are looking for an answer to their search for meaning. More than 30,000 years separate us and yet their works "speak to us" and are now accessible (Carole Fritz 2017).

16.4. The anthropization of the planet

However, this does not mean that humans have totally separated from other living species. They depend on them for food, clothing and to structure their lives. The human being is certainly the only living being to rebuild his environment on a planetary scale, even if he has to adapt to diverse climatic conditions and to the physical and living environment. He must find for himself, even if collectively, the conditions of life in groups that are increasingly great; sometimes there are meetings of human groups that have evolved independently, no longer speak the same language, and live differently. The management of violence, which is always difficult, even

1 These founders of linguistics published between the end of the 19th century and the First World War.

initially among his primate cousins, must be relearned, modified and adapted, according to the growth of the groups.

Even if the last 50 millennia are not homogeneous there are stasis and rapid transformations, analogously in line with the biological evolutionary approach of punctuated equilibria (Gould and Eldredge 1977), the globalization of humanity has been a heavy trend since that *exit from Africa* which seems to have started about 50,000 years ago. The whole planet has been anthropized and technical development has constantly opened up the human horizon until it has gone beyond the limits of the planet. This great movement is associated with the emergence of new technical fields that need to be studied.

Is it always the same movement of innovation and expansion that will continue, or is the human species entering a new, unknown world, for which new capacities, with unknown beginnings, will be mobilized? This is a new field of investigation.

17

The Myth of the Golden Age

The being who is "free" and fully human, in a medium-sized group that understands the world, that invents its own origin and whose basic needs are satisfied thanks to remarkable technical competence, corresponds to an era that seems to have left a lasting mark in the collective memory. The myth of the Golden Age seems to symbolize a mythical and prosperous past; it states that during that age, humans were nomadic, unattached, gathering and hunting, not growing crops or raising livestock. One might think that this is pure nostalgia, constructed at times when life seemed difficult. It is also possible that it corresponds to the difference between childhood and adulthood. For an adult, childhood is often reconstructed as a golden age, either made up of carelessness and innocence or lived at a time that seems better than the present: "things were better before" springs to mind. Still, it is possible that this myth says something about an ancient time.

17.1. The Golden Age in ancient myths

It was Hesiod, in his *Theogony* and in *Works and Days*, who transmitted the myth of an ancient golden age to the West. Ovid takes up this theme again at the beginning of *Metamorphoses*. The Golden Age follows the creation, or appearance, of Man; it is a time of abundance and happiness that suggests a form of innocence. This is anachronistic, because while we know that the Stone Age is an age of abundance, as Sahlins (2017) points out, it cannot be about agriculture and gardens. It is said that "fields produced without culture", so it involved hunting, fishing and gathering. We know that it was not a peaceful age, rather an age of flamboyance, violence and fighting (Guilaine and Zammit 2004; Patou-Mathis 2013). In terms of the population, the number of violent deaths during the Paleolithic era remains high even though population densities are low.

For the Jews, this is written in Genesis; originally humans lived in an earthly paradise where there was no need to work. This can be found in various forms in earlier Chaldean or Sumerian myths. There was an original sin, and, driven out of an otherwise paradisiacal world, if not a very different one, humans had to start to "work by the sweat of their brow" (Roux 1995). Genesis is certainly a copy, but as it seems to be accepted by all monotheisms of Semitic origin and responds to the Greek myth, it must be recognized that it already brings together two thirds of humanity.

For Hindus, the golden age of beginnings is called Satya Yuga (era of purity) and began as humanity emerged in its original state of innocence. The difference with the two previous myths is that the texts that speak of it are later, developing a cyclical conception of history that is highly intellectualized. There is no sin or fall, except to consider the *Mahabharata* as the story of a long fall that tells the end of the Vedic era (hunter-gatherers?) and the establishment of historical times and Kali Yuga, the darkest age. An "objective" fall would have taken place.

It would seem, even if it is not very precise, that there are some references to a fall, which would be a loss of a permanent state of awakening in Taoism and the Western Chan schools, according to a tradition that goes back to the writing of the *Nei King,* one of the books of Chinese antiquity. The *Nei King* describes a golden age, that of an ancient people, the Tchen Jen, whose members lived the Tao to the fullest[1]. Taoist thinkers as well as Confucian thinkers, as early as the 5th Century BCE, refer to a lost golden age. Surprisingly, we find it again in the writings of Carlos Castaneda in his exchanges with his master Don Juan, for whom this fall dates back some 10,000 years. Could it be a trace of some event? Such a late reconstruction may also be a revival of ancient myths, or even worse, a modern reconstruction, a late rationalization combining ancient myths and modern anthropological knowledge.

1 This transcription is written *shengren* in wade, *zhenren* in pinyin, translatable into the Taoist "veritable man"; it comes from Liu Han's *Huainan zi,* completed in 139 B.C. This myth is thus maintained in Taoist thought long afterwards.

17.2. The Golden Age of modern thinkers

It is possible that these myths – and the older and more mythical they are, the more it is imaginable that they say something lived, something existential – recount something that was significant a few millennia ago. Louise O. Fresco (2015) suggests that the transition to agriculture may be what is experienced as the original sin, and that the anguish associated with the construction of this New World had to be calmed. Nostalgia for the Old World could describe this transition. Jean Markale (1999), after Shlomo G. Shoham (2005, pp. 119–121), sets about deconstructing the myth of genesis, while showing that it describes the transition to agriculture and animal husbandry and, over time, to the new form of associated social organization.

The continuation of the biblical myth, about Cain and Abel, describes a bloody struggle, which transmits a certain reality, between breeders and farmers, who are nevertheless brothers, that is, belonging to the same people. The Bible thus clearly states that Adam and Eve are not the first humans (unless we invent that they lived for millennia and had children without ceasing), at best the parents of the first Semites, but we know that they are groups that evolve. This war between breeders and farmers is not terminated; it is visible in recent events, such as the partition of Sudan or the Rwandan genocide, it is not Semitic and seems to be based on two very different logics: that of pastoralism and that of agriculture, which is working the land.

Here, with reference to the optical illusions described above, of the complex reality of human evolution, we look only at the relationship to the technique. The Golden Age, if it is anything other than a nostalgic tendency in a period of change, would correspond to a kind of optimum of a technical age during which technique was made by humans and for humans who made it, even if it is not a question here of idealizing it. The human being could consider himself, even if it was an illusion, as the master of the technique, and maintain companionship with his tools.

Alain Testart (2012) shows that the three major inventions of the Upper Paleolithic, not very inventive, were in close connection with the body: the eye needle, the propeller and the harpoon. In addition, there were a few virtual inventions (ceramics) with no known consequences, he says. He concludes that these three characteristics indicate a low interest in material things. Would he too have been caught up in the myth of a golden age, an age when humans were attentive to the physical body, to being present to oneself, and hardly preoccupied with assets and conquests (Testart 2012, pp. 307–308)?

Marcel Otte (2001), an archaeological prehistorian, is even more explicit. He considers that the Mesolithic period was a moment of liberation from all natural, economic and symbolic constraints, which corresponds to the myth of Lost Paradise, since this is what humans left to cultivate the land. Sahlins, as we have seen, describe this time as an age of plenty. So there are prehistorians who lean towards the possibility of a golden age, before agriculture, before the Neolithic.

It is probable that the human of the Upper Paleolithic, in the context of this new psychic development, in any case largely protected by his technical mastery and by the friendship of the dog, invented the ritualization of the cycles of life, whether it concerns the life of the collective or the life of the individual integrated into the collective, and this at the level of days, seasons, ages, generations, the formation of couples, etc. Through submission to the ritual, something new appeared: the first step towards the possibility of a technical innovation. From the moment when it was possible to coordinate groups and impose rituals, the reference in the name of which it was possible to gather a group could be used to gather groups beyond the number of Dunbar. The human born of the first cognitive revolution was an imaginative being who demanded answers to his questions about the meaning of existence. At the same time, low but real population growth may have led over time to larger community adaptations.

17.3. Believing in a golden age: a cognitive bias?

But we must not forget either that in the search for the sources of this myth, we must add the search for the reasons why it is being proven to the extent that everything that is old ends up falling into a better age. There may be a bias which is to believe that ancient texts tell the truth when they only tell the beliefs of the people of the time. Twenty-six to twenty-eight centuries ago, the estimated date of these writings, we are more than 5,000 years away from the expansion of agriculture in a context of a mostly oral tradition, the common use of local (Sumerian, then Akkadian) and often specialized writing having existed for less than 3,000 years. This reasoning is valid for the other traditions: Greek, Indian, Chinese, Amerindian, etc.

This bias is supported by the fact that the old, in the sense of childhood, is often experienced by an adult as his or her golden age, the rising youth being full of defects and decadent, which has been attested since there have been writings on this subject for more than 26 centuries. This myth of the Golden Age is thus a kind of question mark, but it seems that it is back and may correspond to collective demands.

Conclusion

Post Hominization

The completion of the Paleolithic marked the end of an era that made human. He/She was already a very special animal, surrounded by techniques and rituals. He/She mastered three technical fields: physical, equipped and mental. The question of the apparently contradictory mystery of his/her origin long prevented access to and analysis of the evolutionary process leading to the human being. Myths first came to mind, which for millennia prevented reflective thinking on the subject. It was the coherent unfolding of an immense process over millions of years that could have failed or led to another being. All the evolutions of living beings that have provoked a change known as rupture, or cladogenesis, reveal such an apparent mystery. It all depends on the spatiotemporal scale considered. This evolution towards humankind is specific and this specificity allows us to understand the neotenization of the human and its use of "constitutively anthropological" and "anthropologically constitutive" techniques.

We will always have "missing stages" as we will never have all the successions from generation to generation that go through a process of biological innovation-evolution. How can we collect – and above all find – a copy of the 300 to 400,000 generations since the separation of the hominin and panin lineages? On the other hand, the discovery of sufficient clues and the development of new analysis techniques (in molecular biology) are gradually making it possible to build a coherent succession, all the more so as we have more and more diverse and complementary accounts in this immense world of living beings and the evolutionary lines that have led to human beings.

The riddle of the first bodily technique suggests considering each step separately. It is a form of variability amplified and/or selected by its effect over a long period of genealogical successions and concerning a new form of learning at each stage. It is now known that the parturient female chimpanzee separated from her group and integrated into a human environment does not know how to behave like a mother with her calf if no one is there to show it to her and to guide her; this is learned, late in life.

All higher mammals must learn certain behaviors essential to their survival from their parents or group. Primates, and chimpanzees in particular, have more complex skills because they have developed considerable biological learning skills. Numerous studies confirm that chimpanzees or bonobos, exposed to the human world from an early age, are capable of learning many skills that they seem incapable of without prior exposure. They become more capable of imitating technical gestures.

According to the first hypothesis of this story, verticality was learnt; the first bipeds became vertical as part of a collective pressure to strive towards verticality; this hypothesis is also based on the present observation. In fact, millions of years later, it is still learned, it is not so easy, and it remains fragile, even if the whole biology and morphology of the body, after millions of years, now requires it.

In the film *Ma Loute*, by Bruno Dumont (2016), it is certainly not the beginning of a new decadent century which is described; it is a kind of myth of origins reconstructed in a universe of the early 1900s. Humans have problems with verticality, with elaborated speech, with the use of tools, with the analysis of reasons and objectives; cannibalism does not seem condemned, nor does incest or rape. Actors keep falling, they do not understand the intentions of others, they use techniques imprecisely, dialogues are broken, etc. Love begins to illuminate life and fails.

The question could be: is learning verticality really the process that will be autocatalytic in the long run and from which the retrospective path to what is defined as human is built? This entire text is a long argument that leads to an answer in the affirmative. It is because the common ancestors of bonobos, chimpanzees and humans had a "beginning" of technical learning that verticality, by freeing the horizon and accentuating the role of the hands, is the means "opening the clearing of being", in the Heideggerian sense. In fact, verticality permeates the human imagination and still seems to be an objective that requires undeniable learning to be maintained, even if it is indeed the primary mark of the human being.

It is from this posture that body techniques were learned and, little by little, as much through emulation as imitation, led to the sharing of attention and the development of a permanent tool technique. Voluntary control of breathing, whether it is through fire, wind instruments, songs or words, has even become a method of perceiving the circulation of internal energy. External objects and experiences were shared, exchanged; the teaching of new behaviors was generalized; under the pressure of technical demands, the body evolved, then the brain; the need to understand causes and intentions was selected; from interest in changes of state, which require simple means of communication, hominids moved to interest in causes, intentions and ends, and the need to transmit them in the context of making increasingly complex tools.

Technical processes, both physical and tool-based, are essential factors in the development of these same processes. Their continuous practice provokes a constant selection pressure on the mastery of use, towards techniques that are always more efficient and therefore always more demanding in terms of cognitive capacities. Language can be understood as the end point of a bio-socio-technical evolution of several million years that amplifies cognitive capacities to the point of initiating a process of development of mental techniques in parallel with the needs of communication and exchange. The various human cognitive capacities can be described as mental tools for solving a problem, whether it concerns the manufacture of material tools, adaptation to a new environment, evaluation of predators or game, cooperation or transmission.

In fact, the basic brain structure already existed in higher primates eight million years ago, as well as some remarkable cognitive abilities; it is through an evolutionary process under the pressure of constant selection that superior abilities could be developed, built on the development of new brain areas in strong interaction with the already existing functional spaces; it is what is called "neocortex" that will experience the greatest expansion. Through neuronal plasticity, the human being is extracted from biology; he creates a world of interaction controlling an aspect of biology.

Even if the whole process that leads to the human being is seen from a technical perspective, not everything is technique; technique is a key to understanding the human being. It is one way of approaching humanity; it is a coherent, operational, meaningful representation in an evolutionary approach. The technique gradually makes its way out of the animal without having to call upon a vital force. It is a social construct, which is developed in stages, in co-evolution. Let us note that this construction cannot be understood without integrating desire. Between technique and desire, it is also a relationship of co-evolution. If the object of desire seems so obscure today, it would be because it has been co-constructed with technical differentiation, corporeal first, then equipped, then mental. The object of desire could be the technical practice, per se.

The same goes for the game. If the first bodily technique could be implemented according to a principle of learned and transmitted play, in the long run, it also associates the tool technique. According to Johan Huizinga, *Homo ludens,* the human player, manifests itself particularly in all body games and sports activities. There are many tool techniques used for playing. Thus, play, desire and technique are strongly intertwined and interact with one another.

It might be interesting here to reinterpret Heidegger. He saw the essence of the technique in the *Gestell*, the translation of which is "enframing" and could be "boarding" or "device" according to French translations. It seems that the term "unveiling" could also be interesting. Indeed, according to Heidegger, the technique defines a framework from which nature is revealed; it is a device that "channels" nature. These different translations suggest that, by the technique, nature is seized, almost attached, then summoned, or warned, to reveal itself. This means that Heidegger's Gestell (1958) is not a description of modern technology, nor the distinctive feature of the modern era. As soon as it was introduced, there was a method in the technique that could be described, among other things, as aggressive. The first bodily technique, the mother of all human techniques, can be understood as an aggression against the body, summoned to stand up and respond to collective demands, which is still the case. Yes, the technique is a mode of unveiling that has a form of aggressiveness, in the sense that a desire for mastery is also revealed. The body itself is constrained and questioned; in the long run, it evolves, through a merciless selection that chooses those best suited to technical use. Technique is a slow but merciless conquest. This conquest brings out what allows it to continue to develop and conquer: the incessant search for causes, reasons and motives.

Philosophy pursues this movement, in its systematic and ruthless method, in the sense that what it seeks are causes, reasons and ultimate (or primary) motives. We see here that everything that is affective and emotional, which is already highly evolved in the chimpanzee or the bonobo, has a voice only insofar as it allows us to perceive the motives and sometimes also the reasons. Most often, in the need for affective and emotional differentiation, cognitive processes are called to the rescue. It is indeed cognitive processes, insofar as they apply to data, events, behaviors, perceptions, sensations, emotions, and then words, which allow differentiation and subjective appropriation, or even sublimation of what is analyzed and interpreted.

The brain itself can be understood as a tool that has become at once increasingly complex, increasingly docile and increasingly active; capable of absorbing everything and evolving according to the environment without requiring, for the moment, drastic biological evolution. By offering ever greater possibilities in analysis, memorization, categorization, analogy, imagination, invention, the brain, like nature, reveals itself by transforming itself under the demanding pressure of the technical approach.

Heidegger's reinterpretation erases, in fact and in principle, the idea that it is new. Heidegger believed that the techniques of his time, as he saw them, were different from the old techniques. While it is certain that they do indeed have major differences with those whose phylogenesis we have just analyzed, the fact remains that as far as the *Gestell* is concerned, they have the same characteristics. These are the fundamental techniques that Heidegger is talking about. However, during this "short" moment of a lost paradise magic, they have allowed what, according to Heidegger, modern techniques deprive the human being of: contact with cosmic forces, the feeling of being. Are there any deviations in the development of techniques after the emergence of the human being that would make "monstrous" modern techniques different from previous ones? Since when? What would have been the transformation process? This is what we will try to understand at a later date.

The dynamics of the technical evolution can be understood, from the very beginning of the first tools, as pre-human-technical and then human-technical co-evolution. If the bodily technique is constructed *de facto* with the tool technique, the latter creates a selection pressure in favor of an improvement to the use and then to the manufacture of tools, which will be contingent and dependent in time on the possibilities of biological evolution, because a bodily technique requires a body that responds to it. The constraints of bodily evolution are not those of the evolution of tools. The externalization of techniques began long before *Homo sapiens*. This co-evolution is translated in the long run by an amplification of the body's technicality which can be described as a technical internalization. This new interiority will be able to express itself through language, resulting in an unprecedented expansion in externalization.

In evolutionary terms, this account, if it is accurate, leads to a conclusion that could enlighten our future choices in our postmodern age. Certainly the Darwinian scheme is systematically applied here, but the reader will have noticed that it works, from the biological point of view, "the other way round". The selection pressure imposed by the development of tools leads, step by step, to a new biological species that reverses the relationship with the environment. The selected individual is the technical human, who has developed a bodily technique that only makes sense with a tool environment. As an isolated biological being without tools, he is increasingly deprived. He no longer has the capacity to live in "nature". He goes further: he eliminates little by little and systematically, all the "masters", those at the top of the food chain, "cornerstones" of the ecosystems (felids, canids, ursidae, sharks, killer whales, etc.). He takes their place, but by being insensitive, at least for a certain time, to the disappearance of its prey, because everything is prey for him.

Timothy Taylor has noticed this and insisted on this point: the human being does reverse Darwinism; he causes the selection of the weakest, the weak human who is able to make and use tools, the neogenic being who does not have the capacity to reach a form of autonomy before about 10 years and who will depend on the group in adulthood anyway. In doing so, he eliminates all potential dangers by eradicating dangerous animals and domestication of those that are safe. The loss of certain intrinsic qualities goes in this direction. Miguel Benasayag wonders whether the domestication of the dog has not led to a decline in the human sense of smell. In this case, he applies the "reverse Darwinian scheme". This reverse Darwinian trend is part of the human world, and it is the result of the way it emerged. In this sense, it is not unique to the modern world.

A final aspect that might be interesting to study in more detail would be the question of human characteristics. An article on great apes has shown that great apes are able to read the thoughts of others and to understand when another great ape is wrong, even to the point of helping them (Krupenye *et al.* 2016). Our close cousins show social intelligence, which here has been taken as a starting point for differentiating the lineage of hominins (Buttelman *et al.* 2017). Wouldn't this confirm once again that the fundamental difference lies in this step-by-step constructed complexity that comes from verticalization and the initial technical aptitude, both physical and instrumental, and then mental? Wouldn't the human being be situated in this initial tendency that leads to a process of incessant co-evolution? Are the beginnings of the human being present in the lineage of the *Australopithecus*? Or is it necessary to define the human being by the ability to use fire, or to speak? Or on the contrary, is this question meaningless, because it is a continuum that makes arbitrary the moment defined as the passage to humanity? The human can describe himself as a being in the process of becoming, which is reflected in the long process of hominization and which is also revealed in the periods that follow. This process, which lasted at least seven million years, can be translated afterwards as a partial liberation of biological "locks" in a co-evolutionary logic of technique and of a superior primate who began to learn for the first time a bodily know-how enabling the use of tools transmitted from generation to generation, probably through playing; the transmission of this game became a selection criterion favoring its continuous improvement.

What remains to be done is a narrative focused on the "post-Upper Paleolithic", on the edge of an unknown world, which reveals new aspects of the human being. It reveals them more than it produces them, given the speed of the process. Nevertheless, we can observe a modification in its relationship with technique, among other things by inventing new technical fields. What is being discovered is this general capacity for mental exaptation that comes from the neuronal "plasticity" of the brain, through which new organizations associated with new technical developments will lead to an extraordinary transformation of human communities, to the point that they modify the relationships between humans and lead to the emergence of what we call History. A second cognitive revolution is taking place as a consequence of writing, it seems, from which evolution seems to be accelerating. It also remains to understand the meaning of the technical evolution of the last six centuries before our present time, during which a third cognitive revolution was unveiled that placed humans at the center of the process of the biosphere evolution; this was the entry into a planetary world transformed by humans, which opened up new horizons. What are the links between, on the one hand, the emergence of theories and research on the technique and, on the other hand, the experience of a new unprecedented technical transformation which seems to force us to abandon the human being? On the contrary, perhaps it offers us opportunities to observe and understand what becoming human is.

References

Works referenced in the text

Alsberg, P. (2013). *In Quest of Man: A Biological Approach to the Problem of Man's Place in Nature*. Pergamon, Oxford.

Ameisen, J.-C. (1999). *La sculpture du vivant. Le suicide cellulaire ou la mort créatrice*. Le Seuil, Paris.

Anaximander of Millet. *See* McKirahan R.D. (2010).

Axelsson, E., Ratnakumar, A., Arendt, M.-L., Maqbool, K., Webster, M.T., Perloski, M., Lindblad-Toh, K. (2013). The genomic signature of dog domestication reveals adaptation to a starch-rich diet. *Nature*, 495(7441), 360–364.

Bachelard, G. (2001). *The Formation of the Scientific Mind: A Contribution to a Psychoanalysis of Objective Knowledge*. Clinamen Press Ltd.

Bachelard, G., Ross, A.C.M. (Translator), Frye, N. (Foreword). (1987). *The Psychoanalysis of Fire*. Revised ed. Beacon Press.

Badinter, E. (1982). *Mother love: Myth and Reality*. Macmillan Pub. Co.

Beaumont, P.B. (2011). The Edge: More on Fire-Making by about 1.7 Million Years Ago at Wonderwerk Cave in South Africa. *Current Anthropology*, 52(4), 585–595.

Bednarik, R.G. (2003). Seafaring in the Pleistocene. *Cambridge Archaeological Journal*, 13(1), 41–66.

Bednarik, R.G. (2012). An aetiology of hominin behaviour. *Homo – Journal of Comparative Human Biology*, 63, 319–335.

Bednarik, R.G. (2013). The origins of modern human behavior. In *The psychology of human behavior*, Bednarik, R.G. (ed.). Nova Science Publishers Inc., Commack, 1–58.

Benasayag, M. (2010). *Organismes et artéfacts*. La Découverte, Paris.

Benasayag, M. (2016). *Cerveau augmenté, homme diminué*. La Découverte, Paris.

Berge, C., Ponge, J.-F. (1983). Les caractéristiques du bassin des australopithèques (*A. robustus*, *A. africanus* et *A. afarensis*) sont-elles liées à une bipédie de type humaine ? *Bulletins et Mémoires de la Société d'anthropologie de Paris*, 10(3), 335–354.

Berque, A. (2015). *Écoumène. Introduction à l'étude des milieux humains*. Belin, Paris.

Berthoz, A. (2002), *The Brain's Sense of Movement*. Harvard University Press.

Bickerton, D. (1995). *Language and Human Behaviour*. UCL Press, London.

Blanckaert, C. (2011). Le darwinisme et ses doubles : note sur la linguistique organiciste. *Romantisme*, 154(4), 65–75.

Blasin, D.E., Moran, S., Moisik, S.R., Widmer, P., Dediu, D., Bickel, B. (2019). Human sound systems are shaped by post-Neolithic changes in bite configuration. *Science*, 363, eaav3218.

Boëda, É. (1994). Le concept Levallois : variabilité des méthodes. *Monographie du CRA*, 9.

Boëda, É. (2013). *Technologique et technologie. Une paléohistoire des objets lithiques tranchants*. @rcheo-éditions, Bordeaux.

Bolk, L. (1926). Le problème de la genèse humaine. *Revue française de Psychanalyse*, March–April, 243–279.

Bonneuil, C., Fressoz, J.-B, Fernbach D. (2017). *The Shock of the Anthropocene: The History and Us*. Verso; Reprint edition.

Bourdier, F. (1960). Évolution humaine et néoténie. *Bulletin de la société archéologique et historique de Chelles*, 1(2), 3–21.

Bourg, D. (1996). *L'homme artifice*. Gallimard, Paris.

Bréal, M. (2018). *Semantics. Studies in the Science of Meaning*. Franklin Classics.

Bramble, D.M., Lieberman, D.E. (2004). Endurance running and the evolution of Homo. *Nature,* 432(7015), 345–352.

Bunn, H.T., Gurtov, A.N. (2014). Prey mortality profiles indicate that early Pleistocene *Homo* at Olduvai was an ambush predator. *Quaternary International*, 322(323), 44–53.

Buttelman, D., Buttelmann, F., Carpenter, M., Call, J., Tomasello, M. (2017). Great apes distinguish true from false beliefs in an interactive helping task. *PLoS One*, XII(4).

Caillois, R. (1992). *Les jeux et les hommes : le masque et le vertige*. Gallimard, Paris.

Caillois, R. (ed.) (2001). *Man, Play and Games*. University of Illinois Press.

Cambefort, J.P. (1981). A comparative study of culturally transmitted patterns of feeding habits in the Chacma Baboon *Papio ursinus* and the Vervet Monkey *Cercopithecus aethiops*. *Folia primatology*, 36, 243–263.

Carmody, R.N., Wrangham, R.W. (2009). The energetic significance of cooking. *Journal of Human Evolution*, 57(4), 379–91.

Carus, P. (1893). *The Philosophy of the Tool*. The Open Court Publishing Company, Chicago.

Chabot, P., Hottois, G. (eds) (2003). *Les philosophes et la technique*. Vrin, Paris.

Chaline, J. (2014). *Les horloges du vivant : un nouveau stade de la théorie de l'évolution ?* Hachette littératures, Paris.

Chaline, J. (2006a). *Le singe, l'embryon et l'homme. Une nouvelle clé de lecture de l'histoire de l'homme*. Ellipses, Paris.

Chaline, J. (2006b). *Quoi de neuf depuis Darwin ?* Ellipses, Paris.

Changeux, J.-P. (1985). *Neuronal Man: the Biology of Mind*. Pantheon Books. 1st American edition.

Cohen, C. (1999). *L'Homme des origines*. Le Seuil, Paris.

Cohen, C. (2007). *Un néandertalien dans le métro*. Le Seuil, Paris.

Cohn, J. (1997). How wild wolves became domestic dogs. *BioScience*, 47(11), 725–728.

Cournot, A. (1843). *Exposition de la théorie des chances et des probabilités*. Librairie Hachette, Paris.

Darwin, C. (1882). *The formation of vegetable mould through the action of worms, with observations on their habits*. Lightning Source UK Ltd.

Darwin, C. (2019). *Coral Reefs*. Binker North.

Deacon, T. (1997). *The Symbolic Species*. Penguin Books, London.

Dean, L.G., Kendal, R.L., Shapiro, S.J., Thierry, B., Laland, K.N. (2012). Identification of the social and cognitive processes underlying human cumulative culture. *Science*, 335(60/74), 1114–1118.

Debouzy, M. (1996). La poupée Barbie [Online]. *Clio. Femmes, genre, histoire*, 4. Available at: http://clio.revues.org/446.

Dehaene, S. (2009). *Reading in the Brain: The New Science of How We Read*. Penguin Books.

Delassus, J.-M. (2001). *Le génie du fœtus*. Dunod, Paris.

Delassus, J.-M. (2005). *Psychanalyse de la naissance*. Dunod, Paris.

Deloison, Y. (2013). *Histoire du piéton*. Avenir des sciences, Paris.

Descola, P. (2013). *Beyond Nature and Culture*. University of Chicago Press.

Dessalles, J.-L. (2007). *Why We Talk: The Evolutionary Origins of Language*. Oxford University Press.

Devaux, É. (1921). L'infantilisme de l'homme par rapport aux anthropoïdes et ses conséquences. *Revue des Sciences pures et appliquées*, 32, 276–280.

Diamond, J. (2001). *Le troisième chimpanzé. Essais sur le devenir de l'animal humain*. Gallimard, Paris.

Dortier, J.-F. (2004). *L'homme, cet étrange animal*. Éditions Sciences humaines, Auxerre.

Douze, K. (2011). L'Afrique de l'Est dans la réflexion globale sur le middle stone age. *Annales d'Ethiopie*, 26, 15–51.

Druzhkova, A.S., Thalmann, O., Trifonov, V.A., Leonard, J.A., Vorobieva, N.V., Ovodov, N.D., Graphodatsky, A.S., Wayne, R.K. (2013). Ancient DNA analysis affirms the canid from Altai as a primitive dog. *PLoS ONE*, 8(3), e57754.

Dubois, M.J.F. (2017). *La métaphore et l'improbable. Émergence de l'esprit postscientifique ?* L'Harmattan, Paris.

Dufour, D.-R. (1999). *Lettres sur la nature humaine à l'usage des survivants*. Calmann-Lévy, Paris.

Dufour, D.-R. (2012). *Il était une fois le dernier homme*. Denoël, Paris.

Dunbar, R.I.M. (1992). Neocortex size as a constraint on group size in primates. *Journal of Human Evolution*, 20, 469–493.

Dunbar, R.I.M. (1993a). Coevolution of neocortical size, group size and language in humans. *Behavioral and Brain Sciences*, 16, 681–735.

Dunbar, R.I.M. (1993b). *Theory of Mind and the Evolution of Language*. Cambridge University Press, Cambridge.

Dunbar, R.I.M., Shultz, S. (2007). Evolution in the social brain. *Science*, 317, 1344–1347.

Druelle, F., Berillon, G. (2014). Bipedalism in non-human primates: a comparative review of behavioural and experimental explorations on catarrhines. *BMSAP*, 26 (3–4), 111–120.

Edelman, G. (1990). *Remembered Present: A Biological Theory Of Consciousness*. Basic Books.

Edelman, G. (2004). *Wider Than the Sky: The Phenomenal Gift of Consciousness*. Yale University Press.

Edelman, G., Tononi, G. (2000). *A Universe of Consciousness: How Matter Becomes Imagination*. Odile Jacob, Paris.

Ellul, J. (1964). *The Technological Society*. Translated by John Wilkinson. Knopf, New York. Rev ed. Knopf/Vintage, New York, 1967.

Eren, M.I., Lycett, S.J. (2012). Why Levallois? A morphometric comparison of experimental “preferential” Levallois flakes versus Debitage flakes. *PLoS ONE*, 7(1), e29273.

Ferraro, J.V., Thomas, W., Plummer, T.W., Briana, L., Pobiner, B.L., James, S., Oliver, J.S., Bishop, L.C., Braun, D.R., Peter, W.D., Seaman, J.W., Binetti, K.M., Seaman Jr, J.W., Hertel, F., Potts, R. (2013). Earliest archaeological evidence of persistent hominin carnivory. *PLoS ONE*, 8(4), e62174.

Firth, C., Malafouris, L. (2009). *The Sapient Mind: Archeology Meets Neuroscience*. Oxford University Press, Oxford.

Florio, M., Namba, T., Pääbo, S., Hiller, M., Huttner, W.B. (2016). A single splice site mutation in human-specific ARHGAP11B causes basal progenitor amplification. *Sciences Advances*, 2(12), e1601941.

Florio, M., Borrell, V., Huttner, W.B. (2017). Human-specific genomic signatures of neocortical expansion. *Current Opinion in Neurobiology*, 42, 33–44.

Foucault, M. (1988). *Madness and Civilization: A History of Insanity in the Age of Reason*. Vintage Books.

Foucault, M. (1994). *The Order of Things: An Archaeology of Human Sciences*. Republished in 1994. Ed. Vintage.

Frederick, E. (2019). Humans haven't just changed what dogs look like–we've altered the very structure of their brains. *Science magazine*, 2 September.

Fresco, L., O. (2015). *Hamburgers in Paradise: The Stories behind the Food We Eat*. Princeton University Press.

Fritz, C. (ed.) (2017). *L'art de la préhistoire*. Citadelles et Mazenod, Paris.

Gärdenfors, P. (2003). *How Homo Became Sapiens: On the Evolution of Thinking*. OUP Oxford.

Gärdenfors, P., Högberg, A. (2017). The archaeology of teaching and the evolution of *Homo docens*. *Current Anthropology*, 58(2), 188–208.

Gayon, J., Petit, V. (2019). *Knowledge of Life Today – Conversations on Biology*. ISTE Ltd, London, and Wiley, New York.

Gehlen, A. (2009). *Essais d'anthropologie philosophique*. Maison des sciences de l'Homme, Paris.

Germonpré, M., Sablin, M.V., Stevens, R.E., Hedges, R.E.M., Hofreiter, M., Stiller, M., Després, V.R. (2009). Fossil dogs and wolves from Paleolithic sites in Belgium, the Ukraine and Russia: osteometry, ancient DNA and stable isotopes. *Journal of Archaeological Sciences*, 36, 473–490.

Germonpré, M., Lázničková-Galetová, M., Sablin, M.V. (2012). Paleolithic dog skulls at the Gravettian Předmostí site, the Czech Republic. *Journal of Archaeological Sciences*, 39, 184–202.

Gittelman, R.M., Schraiber, J.G., Vernot, B., Mikacenic, C., Wurfel, M.M., Akey, J.M. (2016). Archaic hominin admixture facilitated adaptation to Out-of-Africa. *Environments. Current Biology*, 26(24), 3375–3382.

Gloor, P.-A. (1986). Néoténie humaine, double puberté, complexe d'Œdipe. *Bulletins et mémoires de la société d'anthropologie de Paris*, 3, XIV(3), 18–1188.

Godelier, M. (2010). *Au fondement des sociétés humaines : ce que nous apprend l'anthropologie*. Flammarion, Paris.

Goethe, J.W. (1808). *Faust I*.

Goldschmidt, R. (1940). *The Material Basis of Evolution*. University Press, New York.

González-Fortes, G., Tassi, F., Trucchi, E., Henneberger, K., Paijmans, J.L.A., Díez-del-Molino, D., Barbujani, G. (2019). A western route of prehistoric human migration from Africa into the Iberian Peninsula. *Proceedings of the Royal Society B: Biological Sciences*, 286, 1895.

Gould, S.J. (1977). *Ontogeny and Phylogeny*. Harvard University Press, Cambridge.

Gould, S.J. (2002). *The Structure of Evolutionary Theory*. Harvard University Press, p. 1464.

Gould, S.J., Eldredge, N. (1977). Punctuated equilibria: the tempo and mode of evolution reconsidered. *Paleobiology*, 3, 115–151.

Gowlett, J.A.J., Harris, J.W.K., Walton, D., Wood, B.A. (1981). Early archaeological sites, hominid remains and traces of fire from Chesowanja, Kenya. *Nature*, 294(12), 125–129.

Granat, J., Peyre, É. (2004). La situation du larynx du genre *Homo*. Données anatomiques, embryologiques et physiologiques. *Biométrie humaine et anthropologie*, 22(3-4), 139–161.

Grof, S., Grof, C. (2010). Holotropic Breathwork: A New Approach to Self-Exploration and Therapy. Excelsior edition.

Groucutt, H.S., Petraglia, M.D., Bailey, G., Scerri, E.M.L., Parton, A., Clark-Balzan, L., Scally, A. (2015). Rethinking the dispersal of *Homo sapiens* out of Africa. *Evolutionary Anthropology: Issues, News, and Reviews*, 24(4), 149–164.

Guchet, X. (2005). *Les sens de l'évolution technique*. Léo Scheer, Paris.

Guilaine, J. and Zammit, J. (2004). *The Origins of War: Violence in Prehistory*. Wiley-Blackwell, Hoboken.

Harari, Y.N. (2015). *Sapiens. A Brief History of Humankind*. Ed. Vintage.

Hare, B. (2002). The domestication of social cognition in dogs. *Science*, 298(5598), 1634–1636.

Hare, B., Tomasello, M. (2005). Human-like social skills in dogs? *Trends in Cognitive Sciences*, 9(9), 439–444.

Harmand, S. *et al.* (2015). 3.3-million-year-old stone tools from Lomekwi 3, West Turkana, Kenya. *Nature*, 521, 310–315.

Hatala, K.G., Demes, B., Richmond, B.G. (2016). Laetoli footprints reveal bipedal gait biomechanics different from those of modern humans and chimpanzees. *Proceedings of the Royal Society B: Biological Sciences*, 283, 1836.

Haudricourt, A.-G., Bert, J.-F. (2010). *Des gestes aux techniques. Essai sur les techniques dans les sociétés pré-machinistes*. Maison des sciences de l'Homme et Quæ, Paris.

Hauser, M.D., Chomsky, N., Fitch Tecumseh, W. (2002). The faculty of language: what is it, who has it, and how did it evolve? *Science*, 298, 156 9–1579.

Havelange, V., Lenay, C., Stewart, J. (2002). Les représentations : mémoire externe et objets techniques. *Intellectica*, 35, 115–131.

Hecht, E.E., Gutman, D.A., Khreisheh, N., Taylor, S.V., Kilner, J., Faisal, A.A., Bradley, B.A., Chaminade, T., Stout, D. (2015). Acquisition of paleolithic toolmaking abilities involves structural remodeling to inferior fronto parietal regions. *Brain Struct. Funct.*, 220, 2315–2331.

Heidegger, M. (2013). *The Question Concerning Technology, and Other Essays.* Harper Perennial Modern Classics. Reissue edition.

Henry, A.G., Brooks, A.S., Piperno, D.R. (2014). Plant foods and the dietary ecology of Neanderthals and early modern humans. *Journal of Human Evolution*, 69, 44–54.

Hessel S, Morin E. (2012). *The Path to Hope.* Other Press.

Hlubik, S. (2018). Finding Prometheus: evidence for fire in the early pleistocene at FXJJ20 AB, Koobi Fora, Kenya. Rutgers.

Hoftstadter D, Sander E. (2013). *Surfaces and Essences: Analogy as the Fuel and Fire of Thinking,* Basic Books.

Hublin, J.-J. (2011). *Quand d'autres hommes peuplaient la Terre.* Flammarion, Paris.

Huizinga, J. (1951). *Homo Ludens. A Study of the Play Element in Culture.* (English edition). Angelico Press.

Indjeian, V.B., Kingman, G.A., Jones, F.C., Guenther, C.A., Grimwood, J., Schmutz, J., Myers, R.M., Kingsley, D.M. (2016). Evolving new skeletal traits by cis-regulatory changes in bone morphogenetic proteins. *Cell,* 164(1-2), 45–56.

Jarrige, F. (2014). *Techno-critique. Du refus des machines à la contestation des technosciences.* La Découverte, Paris.

Jarroson, B. (2010). *Le Temps des magiciens.* Le Pommier, Paris.

Jeusel, É., Mafart, B. (2003). Évolution morphométrique de la cavité palatine des hominidés fossiles. *Anthropologie*, XLI(3), 205–212.

Jeusel, É., Mafart, B. (2006). Le palais des hominidés. Comparaison avec les grands singes et l'homme moderne. Résultats préliminaires. *Antropo*, 11, 235–247.

Jouxtel, P. (2005). *Comment les systèmes pondent. Une introduction à la mémétique.* Le Pommier, Paris.

Kaminski, J. (2004). Word learning in a domestic dog: evidence for "fast mapping". *Science*, 304(5677), 1682–1683.

Kaplan, F. (2001). *Des singes et des hommes, la frontière du langage*. Fayard, Paris.

Kaplan, F. (2008). *L'embryon est-il un être vivant ?* Le Félin, Paris.

Kapp, E. (2007). *Principes d'une philosophie de la technique*. Vrin, Paris.

Kareiva, P., Watts, S., McDonald, R., Bouchet, T. (2007). Domesticated nature: shaping landscapes and ecosystems for human welfare. *Science*, 316(5833), 1866–1869.

Karkanas, P., Shahackgross, R., Ayalon, A., Barmatthews, M., Barkai, R., Frumkin, A., Gopher, A., & Stiner, M. (2007). Evidence for habitual use of fire at the end of the Lower Paleolithic: site-formation processes at Qesem Cave, Israel. *Journal of Human Evolution*, 53(2), 197–212.

Kawai, M. (1965). Newly-acquired pre-cultural behavior of the natural troop of Japanese monkeys on Koshima islet. *Primates*, 6(1), 1–30.

Kellis, M., Wold, B., Snyder, M.P., Hardison, R.C. (2014). Defining functional DNA elements in the human genome. *PNAS*, 111(17), 6131–6138.

Kessel, J. (1959). *The Lion*. Translated by Peter Green. Hart-Davis, London.

Kittler, R., Kayser, M., Stoneking, M. (2003). Molecular evolution of *Pediculus humanus* and the origin of clothing. *Current Biology*, 13(16), 1414–1417.

Krausmann, F., Erb, K.-H., Gingrich, S., Haberl, H., Bondeau, A., Gaube, V., Lauk, C., Plutzar, C., Searchinger, T.D. (2013). *Global human appropriation of net primary production doubled in the 20th Century*. Potsdam Institute of Climate Impact Research, Potsdam.

Krupenye, C., Kano, F., Hirata, S., Call, J., Tomasello, M. (2016). Great apes anticipate that other individuals will act according to false beliefs. *Science*, 354(6308), 110–114.

Kupiec, J.-J. (2012). *L'ontophylogenèse : évolution des espèces et développement de l'individu*. Quae, Versailles.

Lakoff, G, Johnson M. (2003). *Metaphors We Live By*. University of Chicago Press, New edition.

de Lamartine, A. (ed.) (1963). *Harmonies poétiques et religieuses, livre III, II, Milly ou la terre natale*. In *Œuvres complètes*. Gallimard, Paris, 392.

Leakey, M.G., Spoor, F., Dean, M.C., Feibel, C.S., Antón, S.C., Kiarie, C., Leakey, L.N. (2012). New fossils from Koobi Fora in northern Kenya confirm taxonomic diversity in early *Homo*. *Nature*, 488(7410), 201–204.

Lebeau, A. (2006). *L'engrenage de la technique*. Gallimard, Paris.

Leroi-Gourhan, A. (1943). *L'homme et la matière*. Albin Michel, Paris.

Leroi-Gourhan, A. (1945). *Milieu et techniques*. Albin Michel, Paris.

Leroi-Gourhan, A. (2018). *Gesture and speech* (*Le geste et la parole*). MIT Press, Reprint edition. October Books. Albin Michel, Paris.

Levivier, M. (2011). L'hypothèse d'un Homme néoténique comme « grand récit » sous-jacent. *Les sciences de l'éducation. Pour l'ère nouvelle*, 3(44), 77–93.

Lévy-Leblond, J.-M. (1996). *Aux contraires*. Gallimard, Paris.

Lieberman, P. (1984). *The Biology and Evolution of Language*. Harvard University Press, Cambridge.

Lo Sardo, V., Zuccato, C., Gaudenzi, G., Cattane, E. (2012). An evolutionary recent neuroepithelial cell adhesion function of huntingtin implicates ADAM10-Ncadherin. *Nature Neuroscience*, 15, 713–721.

Losey, R.J., Garvie-Lok, S., Leonard, J.A., Katzenberg, M.A., Germonpré, M., Nomokonova, T., Savel'ev, N.A. (2013). Burying dogs in ancient Cis-Baikal, Siberia: temporal trends and relationships with human diet and subsistence practices. *PLoS ONE*, 8(5), e63740.

Lovejoy, C.O. (2009). Reexamining human origins in light of *Ardipithecus ramidus*. *Science*, 326, 74–82.

de Lumley, H. (2017). *La domestication du feu*. Odile Jacob, Paris.

Lyotard, J.-F. Geoff Bennington (Translator), Brian Massumi (Translator), Fredric Jameson (Foreword). (1984). *The Postmodern Condition: A Report on Knowledge*. University Of Minnesota Press.

Malafouris, L. (2013). *How Things Shape the Mind. A theory of material engagement*. MIT Press, Cambridge/London.

Marean, C.W. (2015). Comment *Homo sapiens* a conquis la planète. *Pour la science*, 458, 27–35.

Markale, J. (1999). *The Great Goddess: Reverence of the Divine Feminine from the Paleolithic to the Present*. Inner Tradition.

Masao, F.T., Ichumbaki, E.B., Cherin, M., Barili, A., Boschian, G., Iurino, D.A., Manzi, G. (2016). New footprints from Laetoli (Tanzania) provide evidence for marked body size variation in early hominins. *ELife*, 5.

Maureille, B. (2008). *Qu'est-il arrivé à l'homme de Néandertal ?* Le Pommier, Paris.

Mauss, M. (1936). Les techniques du corps. *Journal de Psychologie*, XXXII, 3–4.

McKirahan, R.D. (2010). *Philosophy before Socrates.* Indianapolis, Hackett Publishing Company Inc., p. 42.

McLean, C.Y., Reno, P.L., Pollen, A.A., Kingsley, D.M. (2011). Human-specific loss of regulatory DNA and the evolution of human-specific traits. *Nature*, 471(7337), 216–219.

McLuhan, M. (2008). *Understanding Media: The Extensions of Man*. Gingko Press.

McLuhan, M. (2011). *The Gutenberg Galaxy* (The Canada 150 Collection). University of Toronto Press.

McPherron, S.P., Alemseged, Z., Marean, C.W., Reed, D., Geraads, D., Bobe, R., Béarat, H.A. (2010). Evidence for stone-tool-assisted consumption of animal tissues before 3.39 million years ago at Dikika, Ethiopia. *Nature*, 466, 857–860.

Mehler, J., Emmanuel Dupoux and Patsy Southgate. (1994). *What Infants Know: The New Cognitive Science of Early Development*. Blackwell Pub.

de Menocal, P.B. (2011). Climate and human evolution. *Science*, 331, 540–542.

Merleau-Ponty, M. (2018). *Phenomenology of Perception.* Franklin Classics, USA.

Meyer, J. (2008). Typology and acoustic strategies of whistled languages: Phonetic comparison and perceptual cues of whistled vowels. *Journal of the International Phonetic Association*, 38(1), 69–94.

Miquel, P.-A. (2015). *Sur le concept de Nature*. Hermann, Paris.

Monod, J. (1997). *Chance and Necessity: Essay on the Natural Philosophy of Modern Biology*. Penguin Press Science.

Morgan, T.J.H., Uomini, N.T., Rendell, L.E., Chouinard-Thuly, L., Street, S.E., Lewis, H.M., Cross, C.P., Evans, C., Kearney, R., de la Torre, I., Whiten, A., Laland, K. (2015). Experimental evidence for the co-evolution of hominin tool-making teaching and language. *Nature communications*, 6, art. 6029.

Morin, E., Piattelli-Palmarini, M. (eds) (1978). *L'unité de l'homme*. Le Seuil, Paris.

Morizot, B. (2011). La néoténie dans la pensée de Gilbert Simondon, ontogenèse d'une hypothèse. In *Cahier Simondon*, 3, Barthélémy, J.-H. (ed.). L'Harmattan, Paris.

Morizot, B. (2016). *Pour une théorie de la rencontre*. Vrin, Paris.

Morris, D. (1967). *The Naked Ape.* Vintage.

Nishimura, T. (2018). The descended larynx and the descending larynx. *Anthropological Science,* 126(1), 3–8.

Nietzsche, F.W. (from 1980 to 2018). *On the Genealogy of Morality or On the Genealogy of Morals.*

Oakley, K. (1949). Man the Tool-Maker. *Bulletin of the British Museum (Natural History).*

Organ, C., Nunn, C.L., Machanda, Z., Wrangham, R.W. (2011). Phylogenetic rate shifts in feeding time during the evolution of Homo. *PNAS*, 108(35), 14555–14559.

Orr, L., Ray, S. (1977). *Rebirthing in the New Age.* Laval, Quebec.

Otte, M. (2001). *Les origines de la pensée. Archéologie de la conscience.* Mardaga, Sprimont.

de Panafieu, J. (1979). *L'illumination intensive.* Retz, Paris.

Patou-Mathis, M. (2013). *Préhistoire de la violence et de la guerre.* Odile Jacob, Paris.

Patterson, N., Richter, D.J., Gnerre, S., Lander, E.S., Reich, D. (2006). Genetic evidence for complex speciation of humans and chimpanzees. *Nature*, 441(7097), 1103–1108.

Perri, A. (2016). A wolf in dog's clothing: initial dog domestication and Pleistocene wolf variation. *Journal of Archaeological Science*, 68, 1–4.

Piaget, J. (1978). *Behavior and Evolution.* Pantheon Books, USA.

Picq, P., Coppens, Y. (2001). *Aux origines de l'humanité. Tome 1. Le propre de l'homme.* Fayard, Paris.

Pinker, S. (2015). *The Language Instinct: How the Mind Creates Language.* Penguin.

Pires, A.E., Detry, C., Chikhi, L., Rasteiro, R., Amorim, I.R., Simões, F., Ginja, C. (2019). The curious case of the Mesolithic Iberian dogs: an archaeogenetic study. *Journal of Archaeological Science*, 105, 116–129.

Pliny the older. (2015). Natural history in thirty-seven books. Book VII. Scholar's Choice Edition.

Pradines, M. (1946). *Traité de psychologie générale. Tome 2. Le génie humain.* PUF, Paris.

Prat, S. (2018). First hominin settlements out of Africa. Tempo and dispersal mode: review and perspectives. *Comptes rendus Palevol* (2017) 1–11.

Prochiantz, A. (2001). *Machine-esprit.* Odile Jacob, Paris.

Prochiantz, A. (2010). *Génétique, évolution, développement.* De Vive Voix, Paris.

Prochiantz, A. (2012). *Qu'est-ce que le vivant ?* Le Seuil, Paris.

Pruetz, J.D., Bertolani, P. (2007). Savanna chimpanzees, *Pan troglodytes verus*, hunt with tools. *Current Biology*, 17(5), 412–417.

Putt, S.S. (2015). The origins of stone tool reduction and the transition to knapping: an experimental approach. *Journal of Archaeological Science: Reports*, 2, 51–60.

Ragir, S. (2000). Diet and food preparation: rethinking early hominin behavior. *Evolutionary Anthropology*, 9, 153–155.

Rank, O. (2015). *The Trauma of Birth*. Routledge, London.

Rawicz, S. (2007). *The Long Walk*. Constable & Robinson, last ed.

Rebato, E., Susanne, C., Brunetto, C. (2003). *Anthropologie biologique. Évolution et biologie humaine*. De Boeck, Brussels.

Reed, D.L., Smith, V.S., Hammond, S.L., Barton, N. (2004). Genetic analysis of lice Supports direct contact between modern and archaic humans. *PLoS Biology*, 2(11), e340.

Reno, P.L. (2014). Genetic and developmental basis for parallel evolution and its significance for hominoid evolution. *Evolutionary Anthropology: Issues, News, and Reviews*, 23(5), 188–200.

Rizzolati, G., Sinigaglia, C. (2010). *Les neurones miroirs*. Odile Jacob, Paris.

Rotjan, R.D., Chabot, J.R., Lewis, S.M. (2010). Social context of shell acquisition in Coenobita clypeatus hermit crabs. *Behavioral Ecology*, 21(3), 639–646.

Roux, G. (1995). *La Mésopotamie*. Le Seuil, Paris.

Ruhlen, M. (1994). *On the Origin of Languages: Studies in Linguistic Taxonomy*. Stanford University Press, Stanford.

Ruhlen, M. (1996). *The Origin of Language: Tracing the evolution of the Mother Tongue*. John Wiley & Sons Inc.

Sahlins, M. (1976). *Culture and Practical Reason*. University of Chicago Press.

Sahlins, M. (2008*). The Western Illusion of Human Nature: With Reflections on the Long History of Hierarchy, Equality and the Sublimation of Anarchy in the West, and ... Conceptions of the Human Condition*. Paradigm.

Sahlins, M. (2017). *Stone Age Economics*. Routledge.

de Saint-Exupéry, A. (2016). *Wind, Sand and Stars*. Penguin Classics.

Sarringhaus, L.A., MacLatchy, L.M., Mitani, J.C. (2014). Locomotor and postural development of wild chimpanzees. *Journal of Human Evolution*, 66, 29–38.

de Saussure, F. (2011). *Course in General Linguistics*. Columbia University Press.

Schäfer, M., Tsekané, S.J., Tchassem, F.A.M., Drakulić, S., Kameni, M., Gonwouo, N.L., & Rödel, M.-O. (2019). Goliath frogs build nests for spawning – the reason for their gigantism? *Journal of Natural History*, 53(21–22), 1263–1276.

Schick, K., Toth, N. (1990). *Making Silent Stones Speak: Human Evolution and the dawn of Technology*. Simon and Schuster, New York.

Schleidt, W.M., Shalter, M.D. (2003). Co-evolution of humans and canids: an alternative view of dog domestication: Homo Homini Lupus? *Evolution and Cognition*, 9(1), 57–72.

Schmid, P. (2004). *Functional Interpretation of the Laetoli Footprints* In D.J. Meldrum & C.E. Hilton (eds), *From Biped to Strider*. Springer, USA.

Seabright, P. (2010). *The Company of Strangers – A Natural History of Economic Life*. Princeton University Press, Princeton, NJ.

Shea, J.J. (2016). *Stone Tools in Human Evolution: Behavioral Differences Among Technological Primates*. Cambridge University Press, Cambridge.

Shoham, S.G. (2005). *Sex as Bait: Eve, Casanova and Don Juan*. Sussex Academic Press.

Sigaut, F. (2012). *Comment Homo devint Faber*. CNRS, Paris.

Simondon, G. (2012). *Two Lessons on Animal and Man*. Univocal Publishing LLC.

Simondon, G. (2020). *Individuation in Light of Notions of Form and Information*. Translated by Taylor Adkins. University of Minesota Press. J. Million, Grenoble.

Simondon, G. (2008). *Imagination et Invention (1965–1966)*. PUF, Paris.

Simondon, G. (2013). *Cours sur la perception*. PUF, Paris.

Simondon, G., Malaspina, C. (translator). (2017). *On the Mode of Existence of Technical Objects*. Univocal, University of Minnesota Press.

Simpson, S.W., Quade, J., Levin, N.E., Butler, R., Dupont-Nivet, G., Everett, M., Semaw, S. (2008). Female *Homo erectus* pelvis from Gona, Ethiopia. *Science*, 322(5904), 1089–1092.

Skoglund, P., Ersmark, E., Palkopoulou, E., Dalén, L. (2015). Ancient wolf genome reveals an early divergence of domestic dog ancestors and admixture into high-latitude breeds. *Current Biology*, 25(11), 1515–1519.

Skoyles, J.R. (2006). Human balance, the evolution of bipedalism and dysequilibrium syndrome. *Medical Hypotheses*, 66(6), 1060–1068.

Sloterdijk, P. (2010). *La domestication de l'être. Pour un éclaircissement de la clairière.* Suivi de *Règles pour le parc humain.* Mille et une nuits, Paris.

Smith, B.P., Litchfield, C.A. (2009). A review of the relationship between Indigenous Australians dingoes (*Canis dingo*) and domestic dogs (*Canis familiaris*). *Anthrozoös*, 22(2), 111–128.

Smyrl, A. (2014). Seen and not heard? Evaluating the prevalence imitation versus guided learning as methods of acquiring skill in the Paleolithic. Thesis, University of Colorado, Boulder.

Snow, C.P. (2018). *The Two Cultures and the Scientific Revolution.* BN Publishing, Hawthrone CA (last ed.).

Sockol, M.D., Raichlen, D.A., Pontzer, H. (2007). Chimpanzee locomotor energetics and the origin of human bipedalism. *Proceedings of the National Academy of Sciences,* 104(30), 12265–12269.

Spoor, F., Wood, B., Zonneveld, F. (1994). Implications of early hominid labyrinthine morphology for evolution of human bipedal locomotion. *Nature,* 369 (6482), 645–648.

Spoor, F., Gunz, P., Neubauer, S., Stelzer, S., Scott, N., Kwekason, A., Dean, M.C. (2015). Reconstructed *Homo habilis* type OH 7 suggests deep-rooted species diversity in early *Homo. Nature*, 519(7541), 83–86.

Stedman, H.H., Kozyak, B.W., Nelson, A., Thesier, D.M., Su, L.T., Low, D.W., Bridges, C.R., Shrager, J.B., Minugh-Purvis, N., Mitchell, M.A. (2004). Myosin gene mutation correlates with anatomical changes in the human lineage. *Nature,* 428(6981), 415–418.

Stiegler, B. (2013). *De la misère symbolique.* Flammarion, Paris.

Stout, D. (2016). Le cerveau à l'âge de pierre. *Cerveau et psycho*, 77, 15–20.

Stout, D., Khreisheh, N. (2015). Skill learning and human brain evolution: an experimental approach. *Cambridge Archeological Journal*, 25, 867–875.

Stringer, C. (2012). *Lone Survivors; How We Came to Be the Only Humans on Earth.* St. Martin's Griffin edition, Reprint (July 30, 2013).

Taylor, T. (2010). *The Artificial Ape: How Technology Changed the Course of Human Evolution.* Palgrave Macmillan, Basingstoke.

Teilhard de Chardin, P. (2008). *The Phenomenon of Man.* Harper Perennial Modern Classics.

Testart, A. (2012). *Avant l'histoire. L'évolution des sociétés, de Lascaux à Carnac.* , Gallimard, Paris.

Thalmann, O., Shapiro, B., Cui, P., Schuenemann, V.J., Sawyer, S.K., Greenfield, D.L., Wayne, R.K. (2013). Complete mitochondrial genomes of ancient canids suggest a european origin of domestic dogs. *Science*, 342(6160), 871–874.

Tinland, F. (1977). *La différence anthropologique. Essai sur les rapports de la nature et de l'artifice.* Aubier Montaigne, Paris.

Tisseron, S. (2015). *Le jour où mon robot m'aimera.* Albin Michel, Paris.

Tomasello, M. (2010). *Origins of Human Communication.* MIT Press.

Trocmé-Fabre, H. (1987). *J'apprends, donc je suis. Introduction à la neuropédagogie.* Éditions d'organisation, Paris.

Varela, F., Thompson, E., Rosch, E. (2017). *The Embodied Mind: Cognitive Science and Human Experience.* MIT Press, revised edition.

von Uexküll, J. (2010). *A Foray into the Worlds of Animals and Humans.* University of Minnesota Press, Minneapolis, MN.

Venkataraman, V.V., Kerby, J.T., Nguyen, N., Ashenafi, Z.T., Fashing, P.J. *et al.* (2015). Solitary Ethiopian wolves increase predation success on rodents when among grazing gelada monkeys herds. *Journal of Mammalogy*, 96, 129–137.

Videan, E.N., & McGrew, W.C. (2002). Bipedality in chimpanzee (Pan troglodytes) and bonobo (Pan paniscus): testing hypotheses on the evolution of bipedalism. *American Journal of Physical Anthropology,* 118(2), 184–190.

Vigarello, G. (2001). *Le corps redressé.* Armand Colin, Paris.

Vilà, C., Savolainen, P., Maldonado, J.E., Amorin, I.R., Honeycutt, R.L., Crandall, K.A., Wayne, R.K. (1997). Multiple and ancient origins of the domestic dog. *Science*, 276(5319), 1687–1689.

Vilà, C., Maldonado, J.E., Wayne, R.K. (1999). Phylogenetic relationships, evolution, and genetic diversity of the domestic dog. *Journal of Heredity*, 90(1), 71–77.

de Waal, F. (2005). *Our Inner Ape.* Riverhead Books, New York, last ed.

de Waal, F. (2009). *The Age of Empathy: Nature's Lessons for a Kinder Society* (English Edition). Crown.

de Waal, F. (2014). *The Bonobo and the Atheist: In Search of Humanism Among the Primates.* W. W. Norton & Company, Reprint edition.

Weil, S. (2001). *Oppression and Liberty.* Routledge Classics.

Winnicott, D.W. (2005). *Playing and Reality.* Routledge, London.

Wood, B. (2014). Le buissonnant rameau. *Pour la Science*, 445, 28–36.

Wrangham, R. (2017). Control of fire in the Paleolithic: evaluating the cooking hypothesis. *Current Anthropology*, 58(S16), 303–313.

Wrangham, R., Carmody, R. (2010). Human adaptation to the control of fire. *Evolutionary Anthropology: Issues, News, and Reviews*, 19(5), 187–199.

Yamagiwa, J. (2010). Research History of Japanese Macaques in Japan. In *The Japanese Macaques*, Nakagawa, N., Nakamichi, M., Sugiura, H. (eds). Springer-Verlag, Tokyo.

Zink, K.D., Lieberman, D.E. (2016). Impact of meat and Lower Palaeolithic food processing techniques on chewing in humans. *Nature*, 531(7595), 500–503.

Zink, K.D., Lieberman, D.E., Lucas, P.W. (2014). Food material properties and early hominin processing techniques. *Journal of Human Evolution*, 77, 155–166.

Some background works not referenced in the text

Aiello, L.C. (1996). Terrestriality, bipedalism and the origin of language. *Proceedings of the British Academy*, 88, 269–289.

Basquiast, J.-P. (2010). *Le paradoxe du* sapiens. *Êtres technologiques et catastrophes annoncées.* J.-P. Bayol, Alès.

de Beer, G.R. (1930). *Embryology and Evolution*. Oxford University Press, Oxford.

de Beer, G.R. (1940). *Embryos and Ancestors*. Oxford University Press, Oxford.

Bourguignon, A. (1989). *Histoire naturelle de l'homme. Tome 1. L'homme imprévu*. PUF, Paris.

Bourguignon, A. (1994). *Histoire naturelle de l'homme. Tome 2. L'homme fou*. PUF, Paris.

Cyrulnik, B. (1983). *Mémoire de singe et paroles d'homme*. Hachette, Paris.

Danchin, A. (1983). *L'œuf et la poule. Histoire du code génétique*. Fayard, Paris.

Demoule, J.-P., Garcia, D., Schnapp, A. (eds) (2018). *Une histoire des civilisations. Comment l'archéologie bouleverse nos connaissances*. La Découverte/Inrap, Paris.

Dennett, D.C. (1990). *La stratégie de l'interprète*. Gallimard, Paris.

Dennett, D.C. (2004). *Théorie évolutionniste de la liberté*. Odile Jacob, Paris.

Dewey, J. (2016). *L'influence de Darwin sur la philosophie*. Gallimard, Paris.

Dobzhansky, T. (1966). *L'homme en évolution*. Flammarion, Paris.

Dupuy, J.-P. (2002). *Pour un catastrophisme éclairé*. Le Seuil, Paris.

Gallagher, S. (2006). *How the Body Shapes the Mind*. Clarendon Press, Oxford.

Gallien, C.-L. (1998). *Homo. Histoire plurielle d'un genre très singulier*. PUF, Paris.

Girard, R. (1978). *Des choses cachées depuis la fondation du monde*. Grasset, Paris.

Guchet, X. (2010). *Pour un humanisme technologique*. PUF, Paris.

Hagège, C. (1985). *L'homme de paroles*. Fayard, Paris.

Haldane, J.B.S. (1932). *The Causes of Evolution*. Longmans, Green and Co. and Harper Brothers, London/New York.

Meillassoux, Q. (2006). *Après la finitude. Essai sur la nécessité de la contingence*. Le Seuil, Paris.

Picq, P., Serres, M., Vincent, J.-D. (2004). *Qu'est-ce que l'humain ?* Le Pommier, Paris.

Reichholf, J. (1991). *L'émergence de l'homme*. Flammarion, Paris.

Rosset, C. (1973). *L'anti-nature*. PUF, Paris.

Sahnouni, M. (2005). *Le paléolithique en Afrique. L'histoire la plus longue*. Artcom'/Errance, Paris.

Tattersall, I. (1999). *L'émergence de l'homme. Essai sur l'évolution et l'unicité humaine*. Gallimard, Paris.

Wilson, E.O. (2003). *The Future of Life. Vintage*. Reprint edition.

Wright, R. (1995). *The Moral Animal: Why We Are the Way We Are: The New Science of Evolutionary Psychology*. Vintage, Reprint edition.

Index

A

B

C

J, K

L

M

N

O

P

V, W

Other titles from

iSTE

in

Science, Society and New Technologies

2020

BARNOUIN Jacques
The World's Construction Mechanism: Trajectories, Imbalances and the Future of Societies
(Interdisciplinarity between Biological Sciences and Social Sciences Set – Volume 4)

ETCHEVERRIA Olivier
The Restaurant, A Geographical Approach: From Invention to Gourmet Tourist Destinations
(Tourism and Mobility Systems Set – Volume 3)

JEANNERET Yves
The Trace Factory
(Traces Set – Volume 3)

MARTI Caroline
Cultural Mediations of Brands: Unadvertization and Quest for Authority
(Communication Approaches to Commercial Mediation SET – Volume 1)

SCHMITT Daniel, THÉBAULT Marine, BURCZYKOWSKI Ludovic
Image Beyond the Screen: Projection Mapping

VIOLIER Philippe, with the collaboration of TAUNAY Benjamin
The Tourist Places of the World
(Tourism and Mobility Systems Set – Volume 2)

2019

BRIANÇON Muriel
The Meaning of Otherness in Education: Stakes, Forms, Process, Thoughts and Transfers
(Education Set – Volume 3)

DESCHAMPS Jacqueline
Mediation: A Concept for Information and Communication Sciences
(Concepts to Conceive 21st Century Society Set – Volume 1)

DOUSSET Laurent, PARK Sejin, GUILLE-ESCURET Georges
Kinship, Ecology and History: Renewal of Conjunctures
(Interdisciplinarity between Biological Sciences and Social Sciences Set – Volume 3)

DUPONT Olivier
Power
(Concepts to Conceive 21st Century Society Set – Volume 2)

FERRARATO Coline
Prospective Philosophy of Software: A Simondonian Study

GUAAYBESS Tourya
The Media in Arab Countries: From Development Theories to Cooperation Policies

HAGÈGE Hélène
Education for Responsibility
(Education Set – Volume 4)

LARDELLIER Pascal
The Ritual Institution of Society
(Traces Set – Volume 2)

LARROCHE Valérie
The Dispositif
(Concepts to Conceive 21st Century Society Set – Volume 3)

LATERRASSE Jean
Transport and Town Planning: The City in Search of Sustainable Development

LENOIR Virgil Cristian
Ethically Structured Processes
(Innovation and Responsibility Set – Volume 4)

LOPEZ Fanny, PELLEGRINO Margot, COUTARD Olivier
Local Energy Autonomy: Spaces, Scales, Politics
(Urban Engineering Set – Volume 1)

METZGER Jean-Paul
Discourse: A Concept for Information and Communication Sciences
(Concepts to Conceive 21st Century Society Set – Volume 4)

MICHA Irini, VAIOU Dina
Alternative Takes to the City
(Engineering, Energy and Architecture Set – Volume 5)

PÉLISSIER Chrysta
Learner Support in Online Learning Environments

PIETTE Albert
Theoretical Anthropology or How to Observe a Human Being
(Research, Innovative Theories and Methods in SSH Set – Volume 1)

PIRIOU Jérôme
The Tourist Region: A Co-Construction of Tourism Stakeholders
(Tourism and Mobility Systems Set – Volume 1)

PUMAIN Denise
Geographical Modeling: Cities and Territories
(Modeling Methodologies in Social Sciences Set – Volume 2)

WALDECK Roger
Methods and Interdisciplinarity
(Modeling Methodologies in Social Sciences Set – Volume 1)

2018

BARTHES Angela, CHAMPOLLION Pierre, ALPE Yves
Evolutions of the Complex Relationship Between Education and Territories
(Education Set – Volume 1)

BÉRANGER Jérôme
The Algorithmic Code of Ethics: Ethics at the Bedside of the Digital Revolution
(Technological Prospects and Social Applications Set – Volume 2)

DUGUÉ Bernard
Time, Emergences and Communications
(Engineering, Energy and Architecture Set – Volume 4)

GEORGANTOPOULOU Christina G., GEORGANTOPOULOS George A.
Fluid Mechanics in Channel, Pipe and Aerodynamic Design Geometries 1
(Engineering, Energy and Architecture Set – Volume 2)

GEORGANTOPOULOU Christina G., GEORGANTOPOULOS George A.
Fluid Mechanics in Channel, Pipe and Aerodynamic Design Geometries 2
(Engineering, Energy and Architecture Set – Volume 3)

GUILLE-ESCURET Georges
Social Structures and Natural Systems: Is a Scientific Assemblage Workable?
(Social Interdisciplinarity Set – Volume 2)

LARINI Michel, BARTHES Angela
Quantitative and Statistical Data in Education: From Data Collection to Data Processing
(Education Set – Volume 2)

LELEU-MERVIEL Sylvie
Informational Tracking
(Traces Set – Volume 1)

SALGUES Bruno
Society 5.0: Industry of the Future, Technologies, Methods and Tools
(Technological Prospects and Social Applications Set – Volume 1)

TRESTINI Marc
Modeling of Next Generation Digital Learning Environments: Complex Systems Theory

2017

ANICHINI Giulia, CARRARO Flavia, GESLIN Philippe, GUILLE-ESCURET Georges
Technicity vs Scientificity – Complementarities and Rivalries (Interdisciplinarity between Biological Sciences and Social Sciences Set – Volume 2)

DUGUÉ Bernard
Information and the World Stage – From Philosophy to Science, the World of Forms and Communications (Engineering, Energy and Architecture Set – Volume 1)

GESLIN Philippe
Inside Anthropotechnology – User and Culture Centered Experience (Social Interdisciplinarity Set – Volume 1)

GORIA Stéphane
Methods and Tools for Creative Competitive Intelligence

KEMBELLEC Gérald, BROUDOUS EVELYNE
Reading and Writing Knowledge in Scientific Communities: Digital Humanities and Knowledge Construction

MAESSCHALCK Marc
Reflexive Governance for Research and Innovative Knowledge (Responsible Research and Innovation Set - Volume 6)

PARK Sejin, GUILLE-ESCURET Georges
Sociobiology vs Socioecology: Consequences of an Unraveling Debate (Interdisciplinarity between Biological Sciences and Social Sciences Set – Volume 1)

PELLÉ Sophie
Business, Innovation and Responsibility (Responsible Research and Innovation Set – Volume 7)

2016

BRONNER Gérald
Belief and Misbelief Asymmetry on the Internet

EL FALLAH SEGHROUCHNI Amal, ISHIKAWA Fuyuki, HÉRAULT Laurent, TOKUDA Hideyuki
Enablers for Smart Cities

GIANNI Robert
Responsibility and Freedom
(Responsible Research and Innovation Set – Volume 2)

GRUNWALD Armin
The Hermeneutic Side of Responsible Research and Innovation
(Responsible Research and Innovation Set – Volume 5)

LAGRAÑA Fernando
E-mail and Behavioral Changes: Uses and Misuses of Electronic Communications

LENOIR Virgil Cristian
Ethical Efficiency: Responsibility and Contingency
(Responsible Research and Innovation Set – Volume 1)

MAESSCHALCK Marc
Reflexive Governance for Research and Innovative Knowledge
(Responsible Research and Innovation Set – Volume 6)

PELLÉ Sophie, REBER Bernard
From Ethical Review to Responsible Research and Innovation
(Responsible Research and Innovation Set – Volume 3)

REBER Bernard
Precautionary Principle, Pluralism and Deliberation: Sciences and Ethics
(Responsible Research and Innovation Set – Volume 4)

VENTRE Daniel
Information Warfare – 2nd edition